WIRELESS CMOS FREQUENCY SYNTHESIZER DESIGN

WIRELESS CMOS FREQUENCY SYNTHESIZER DESIGN

by

J. Craninckx
Alcatel Mietec

and

M. Steyaert
Katholieke Universiteit Leuven

KLUWER ACADEMIC PUBLISHERS
BOSTON / DORDRECHT / LONDON

A C.I.P. Catalogue record for this book is available from the Library of Congress.

ISBN 0-7923-8138-6

Published by Kluwer Academic Publishers,
P.O. Box 17, 3300 AA Dordrecht, The Netherlands.

Sold and distributed in the North, Central and South America
by Kluwer Academic Publishers,
101 Philip Drive, Norwell, MA 02061, U.S.A.

In all other countries, sold and distributed
by Kluwer Academic Publishers,
P.O. Box 322, 3300 AH Dordrecht, The Netherlands.

Printed on acid-free paper

Printed in the Netherlands.

Contents

List of Figures

List of Tables

List of Symbols and Abbreviations

Symbols

α	Transconductor safety margin
β	FM modulation index
δ	Horizontal bend in bonding wire Skin depth
ε	PLL settling accuracy
ε_{ox}	Permittivity of oxide
ρ	Conductivity
ρ_{Au}	Conductivity of gold
ρ_{epi}	Conductivity of epi-layer
ρ_{subs}	Conductivity of bulk substrate
θ	Phase
$\theta(t)$	Phase (time-varying)
θ_{div}	Frequency divider output phase
θ_{err}	Phase error
$\theta_{err,ss}$	Steady-state phase error
θ_{out}	VCO output phase
θ_p	Peak phase deviation
θ_{ref}	Reference phase
$\sigma_{\Delta\tau}$	Timing error variance
$\Delta\tau$	Timing error
τ_c	Time constant of ω_c
τ_p	Time constant of ω_p
τ_z	Time constant of ω_z
μ	Permeability
μ_{Au}	Permeability of gold

$\Delta\omega$	Offset frequency
$\Delta\omega_L$	Lock-in range
$\Delta\omega_P$	Pull-in range
$\Delta\omega_H$	Hold-in range
$\Delta\omega_{1/f^3}$	$\Delta\omega^{-3}$ corner frequency
$\delta(\omega)$	Dirac impulse function
ω	Frequency $[rad/s]$
ω_0	Resonance frequency, oscillation frequency
ω_a	Loop filter low-frequency pole frequency
ω_c	PLL cross-over frequency
ω_m	Modulating (baseband) frequency
ω_n	Natural frequency
ω_p	Loop filter pole frequency Parallel resonance frequency
ω_s	Series resonance frequency
ω_z	Loop filter compensating zero frequency
ζ	Damping factor
A	Amplitude Amplifier noise contribution factor
$A(t)$	Amplitude (time-varying)
A_i	Amplitude of signal i
c_π	Bipolar small-signal base-emittor capacitance
C	Capacitance
C_b	Bridging capacitance
C_c	Crystal's parallel capacitance
C_{eff}	Effective capacitance
C_L	Gyrator load capacitance
C_p	Capacitance that sets ω_p Parallel Capacitance Parasitic Capacitance
C_{pad}	Bonding pad parasitic capacitance
C_s	Series Capacitance
C_z	Capacitance that sets ω_z
d	Distance between bonding wires
Di	Diode no. i
di_X^2	Noise current of element X
dv_X^2	Noise voltage of element X
dV_{out}^2	Oscillator output noise voltage
dV_{out,R_x}^2	Output noise voltage for parasitic resistor R_x
dBc	deciBell carrier : power relative to the carrier power
E	Energy

f, F	Frequency $[Hz]$
F	Excess noise factor
f_0	Resonance frequency, oscillation frequency
f_T	Transistor unity current gain frequency
F_X	Excess noise factor of element X
F_{div}	Divided frequency, i.e. output frequency of a frequency divider
F_{in}	Input frequency
F_{out}	Output frequency
F_{ref}	Reference frequency of a PLL
$G(s)$	Forward transfer function
g_m, G_M	Transconductance
g_o	Transistor small-signal output conductance
$G_{M,L}$	Total active inductor transconductance
G_{M,R_x}	Oscillator transconductance for parasitic resistor R_x
$G_{lf}(s)$	Loop filter transfer function
$GH(s)$	Open loop transfer function
h	Height of vertical bend in bonding wire
$H(s)$	Feedback factor
$H_{noise,R_x}(s)$	Inverse of $T_{noise,R_x}(s)$
I	Current
Ii	Current source no. i
I_p	Charge pump current
I_C	Charge pump output current
K_F	PLL forward gain $[s^{-1}]$
J_i	i^{th} order Bessel function of the first kind
K_{lf}	Loop filter gain
K_{pd}	Phase detector gain factor $[V/rad]$
K_{vco}	VCO gain factor $[rad/Vs]$
ℓ	Length of bonding wire
L	Inductance
L_{act}	Active inductance
L_{tot}	Total inductance
L_s	Series inductance
$\mathcal{L}$	Single-sided phase noise spectral density
M	Active inductor noise contribution factor Mutual inductance
Mi	MOS transistor no. i
N	Integer division modulus
N_p	Prescaler integer division modulus
n	Number of bits Fractional ($0 \leq n \leq 1$) division modulus

	Bulk modulation noise factor
P	Preset number
	Power
Q	Quality factor
Q_L	loaded quality factor
Q_{R_x}	Quality factor due to parasitic resistor R_x
Qi	Bipolar transistor no. i
r	Radius of bonding wire cross section
r_π	Bipolar small-signal base-emittor resistance
r_b	Bipolar small-signal base resistance
r_C	Capacitance ratio
Ri	Resistor no. i
$R_{bondwire}$	Equivalent bonding wire resistance due to skin effect losses
R_c	Parasitic series resistance of a capacitor
R_{eff}	Effective resistance of an LC-tank
R_{epi}	Resistance of epi-layer
R_l	Parasitic series resistance of an inductor
R_m	Magnetic losses resistance
R_n	Equivalent noisy resistance
R_p	Resistance that sets ω_p
	Parallel resistance
R_{pad}	Bonding pad parasitic resistance
R_s	Series resistance
R_{subs}	Resistance of bulk substrate
$R_{substrate}$	Equivalent bonding wire resistance due to substrate losses
R_{skin}	High-frequency resistance of bonding wire
R_{wire}	DC resistance of bonding wire
R_z	Resistance that sets ω_z
S	Swallow counter division factor
S_θ	Phase power spectral density
$S_{\Delta\omega}$	Instantaneous frequency deviation power spectral density
$S_{V_{out}}$	Output voltage power spectral density
$T_{loop,R_x}(s)$	Loop transfer function with parasitic resistor R_x
$T_{noise,R_x}(s)$	Noise transfer function for parasitic resistor R_x
T_d	Delay time
T_p	Average period time
T_ε	PLL settling time
V	Voltage
V_A	Voltage amplitude
v_c, V_c	VCO control voltage
V_{eff}	Effective voltage signal

V_{int}	Internal voltage
V_{out}	Output voltage
v_{pd}	Phase detector output voltage
Y	Admittance
Z_{lf}	Charge pump loop filter impedance

Abbreviations

AC	Alternating Current
ADC	Analog-to-Digital Converter
AGC	Automatic Gain Control
AM	Amplitude Modulation
AMPS	American Mobile Phone System
AT&T	American Telephone and Telegraph company
BiCMOS	Bipolar and CMOS
BS	Base Station
BSC	Base Station Controller
BST	Base Station Transceiver
BW	Bandwidth
CCO	Current-Controlled Oscillator
CEPT	Conférence Europeénne des Postes et Télécommunications
CMOS	Complementary Metal Oxide Semiconductor
DAC	Digital-to-Analog converter
DC	Direct Current
DDFS	Direct Digital Frequency Synthesis
DMP	Dual-Modulus Prescaler
DRO	Dielectric Resonance Oscillator
DUT	Device Under Test
ETSI	European Telecommunications Standards Institute
EXOR	Exlusive OR
FCC	United States Federal Communication Commission
FFT	Fast Fourier Transform
GaAs	Gallium Arsenide
GSM	Groupe Spécial Mobile / Global System for Mobile communications
HLR	Home Location Register
IC	Integrated Circuit
IF	Intermediate Frequency
LO	Local Oscillator
LPF	Low-Pass filter
MS	Mobile Station

MSC	Mobile Service and switching Center
MTX	Mobile Telephone Exchange
NAND	Negative AND
NMOS	N-channel Metal Oxide Semiconductor
NMT	Nordic Mobile Telephony group
NSS	Network and Switching Subsystem
PD	Phase Detector
PFD	Phase-Frequency Detector
PSD	Power Spectral Density
PHP	Personal Handy Phone
PLL	Phase-Locked Loop
PM	Phase Modulation
PMOS	P-channel Metal Oxide Semiconductor
PTT	Post, Telephone and Telegraph
RF	Radio Frequency
SAW	Surface Acoustic Wave filter
SNR	Signal-to-Noise Ratio
SSB	Single SideBand
TCXO	Temperature-Controlled Crystal Oscillator
VCO	Voltage-Controlled Oscillator
VCXO	Voltage-Controlled Crystal Oscillator
VLR	Visitor Location Register
YIG	Yttrium Iron Garnite

Bibliographic Conventions

In this text, bibliographic references contain information of the first author, the publication source and the year of publication, poosibly extended with an extra character when more than one publication of the same author has been published in the same journal in the same year. In this way the reader already finds a lot of information on bibliographic references within the text. The full information can of course be found in the bibliography after the last chapter. An example is [Crani ISSCC95]. "Crani" are the first five letters of the first author's name, i.e. Craninckx. "ISSCC" is an abbreviation forthe journal of conference the reference was published in. "95" is the year of publication. If the naming scheme is not sufficient for uniquely refer to a work, a letter (a, b, etc.) will be added to avoid ambiguity with another bibliographic reference. A list of the most important abbreviations used for pubication sources is given below. The absence of such an abbreviation in the reference indicates that it refers to a book.

AACD	Proceedings of the workshop Advances in Analog Circuit Design
ACD	Analog Circuit Design

AICSP	Analog Integrated Circuits and Signal Processing journal
CAS	IEEE Transactions on Circuits and Systems
CASI	IEEE Transactions on Circuits and Systems I : Fundamental Theory and Applications
CASII	IEEE Transactions on Circuits and Systems II : Analog and Digital Signal Processing
CICC	Proceedings of the Custom Integrated Circuits Conference
EL	IEE Electronic Letters
ESSC	Proceedings of the European Solid-State Circuits Conference
IEEE	Proceedings of the IEEE
ISCAS	Proceedings of the International Symposium on Circuits and Systems
ISSCC	Digest of technical papers of the International Solid-State Circuits Conference
JSSC	IEEE Journal of Solid-State Circuits
VLSI	Proceedings of the Symposium on VLSI Circuits

Preface

The recent boom in the mobile telecommunication market has trapped the interest of almost all electronic and communication companies worldwide. New applications arise every day, more and more countries are covered by digital cellular systems and the competition between the several providers has caused prices to drop rapidly. The creation of this essentially new market would not have been possible without the appearance of small, low-power, high-performant and certainly low-cost mobile terminals. The evolution in microelectronics has played a dominant role in this by creating digital signal processing (DSP) chips with more and more computing power and combining the discrete components of the RF front-end on a few ICs. This work is situated in this last area, i.e. the study of the full integration of the RF transceiver on a single die. Furthermore, in order to be compatible with the digital processing technology, a standard CMOS process without tuning, trimming or post-processing steps must be used. This should flatten the road towards the ultimate goal : the single chip mobile phone. The local oscillator (LO) frequency synthesizer poses some major problems for integration and is the subject of this work.

The first, and also the largest, part of this text discusses the design of the Voltage-Controlled Oscillator (VCO). The general phase noise theory of LC-oscillators is presented, and the concept of effective resistance and capacitance is introduced to characterize and compare the performance of different LC-tanks. Enhanced LC-tanks are presented which use a combination of several inductors and capacitors to allow a trade-off between phase noise and power consumption.

Two possible implementations of integrated inductors are investigated in great detail. The first type uses the parasitic inductance usually associated with a bonding wire to its advantage. The low series resistance and small parasitic capacitance lead to extremely high quality factors and hence the possibility of low phase noise. A 1.8-GHz implementation achieves indeed state-of-the-art phase noise performance. The more standard types of integrated inductors are spiral coils laid out in the metal routing levels. They are characterized completely using efficient finite-element simulation techniques, in order to understand all the losses present. Some design guidelines are presented, the most important of which is the use of a hollow coil to avoid eddy current losses in the inner conductor tracks. The VCO designs presented indeed achieve a large improvement in phase noise performance over comparable designs, which must be totally contributed to the optimization of the inductor coil.

The second part of this work discusses the other high-frequency building block of the PLL, i.e. the frequency divider or prescaler. A new dual-modulus prescaler operating principle, the phase-switching architecture, is presented which allows the realization of more than one division modulus without sacrificing speed with respect to a fixed-modulus divider. Two submicron designs, one of which implements eight division moduli, demonstrate the capabilities of this prescaler architecture in the GHz frequency range.

Finally, both building blocks are combined together with the low-frequency parts of a PLL in a fourth-order type-2 charge pump PLL frequency synthesizer that aims at the DCS-1800 application. The loop filter also uses no external elements and is implemented in a dual signal path configuration to limit the size of the integrated capacitors. A special linearization technique ensures a constant open loop gain and hence sufficient stability over the full VCO frequency range. This prototype design demonstrates the feasibility of a standard CMOS integrated GHz frequency synthesizer, and thereby brings us one step closer to a full CMOS RF transceiver.

We also wish to express our gratitude to all persons who have contributed to the realization of this book and to the research described in it. We would like to thank Prof. W. Sansen and Prof. B. Nauwelaerts for carefully proofreading the manuscript. Our thanks also goes to the NFWO (The Belgium National Fund for Scientific Research) for funding of the research, and all our collegues at the ESAT-MICAS laboratories for their direct or indirect contributions to this work.

Finally, we thank our families for their support and patience. Without it this research and this book would not have been possible.

Jan Craninckx
Michiel Steyaert

Department of Electrical Engineering - ESAT MICAS
Katholieke Universiteit Leuven
Leuven, Belgium, 1997

1 INTRODUCTION

1.1 MODERN TELECOMMUNICATIONS

If every decade has its name, than the nineties are certainly the decade of mobile communication. Never before the telecommunication market has changed so much. This has been made possible by major changes in two areas. First, there is the political aspect. For several decades, every country had its own PTT (Post, Telephone and Telegraph) institute, a government-owned company that enjoyed a monopoly to exploit the country's fixed telephone system. Mobile telephony services were provided by these PTTs, and were regarded as an addition to the fixed network. The liberalization imposed by the European Union will put an end to this by the end of 1998. Most PTTs have already gone public, and in a lot of countries there is already at least one competitor operating a mobile network. Competition between the operators will certainly cause a drop in prices, leading to an increase in mobile users.

Secondly, there is the technology evolution. Or should it be revolution ? During the eighties, the large, bulky analog mobile phones only offered low-quality conversations, which were interrupted at the most irritating moments, and owning one was very expensive. Nowadays, a GSM (Groupe Spécial Mobile, or Global System for Mobile communications) phone fits into your pocket, offers you a high-quality connection and several hours of talk time, and above all, has become a lot cheaper than before.

1.1.1 A Short History

It may surprise many people that last year, in 1996, mobile telephony celebrated its fiftieth birthday. The possibilities of radio communications were first demonstrated by Guglielmo Marconi in the mid-1890s. In 1901, he managed to send a signal across the Atlantic Ocean. The first uses were for telegraph messages, but in 1906 the American Reginald Fessenden was able to transmit the human voice successfully by radio. During the '20s and '30s, the radio receiver came into everybody's home, as a new and more-or-less bulky piece of furniture. And it was in 1946, more than 50 years ago, that AT&T obtained approval from the US Federal Communications Commission (FCC) to operate the world's first commercial car-borne, or mobile, telephone service. This was in St Louis, Missouri. The system had a single base station, on high ground, and was equipped with six channels. It was a big success and within a year, AT&T had mobile networks serving 25 other cities [Muerl 94].

These systems were based on a very simple principle : find the highest building in the area, set up the biggest antenna you can manage, and transmit as much power as you can afford. With luck, you will have covered most of your target area. A big problem was of course the power transmitted by the mobile station. A large extra battery and a powerful generator were required. Another problem was the available frequency spectrum, which limited the number of users.

The answer to these problems was presented by Bell Labs as early as 1947 : the cellular concept. The total area to be covered is divided in smaller areas - cells - each with its own base station using a number of frequencies. Cells spaced far apart can re-use the same frequencies, and the transmitted power is much smaller due to the small cell area. The "hand-off" function allows to switch from one cell to another during a call in progress when a car crosses a cell boundary. This proposal was presented to the FCC, but since the critical technology for the hand-off function was not yet available, nothing further happened. Twenty-one years later, in 1968, the FCC initiated proceedings. This was the first go-ahead for cellular, but also the beginning of several years of lobbying and counter-lobbying between the several competitors, resulting finally in 1981 in a 40-MHz allocation for two licenses in each market, where a market is defined as a geographical area, such as a city. Finally, on October 13th 1983, the first commercial cellular system in the US began operation in Chicago.

These 36 years of development seem very long. They were partly due to development of necessary key technology functions. The transistor was invented in 1948, and started the miniaturization of the mobile phones. The microprocessor enabled to handle the hand-off function between cells with very little disruption of the conversation. Digital computers controlling analog switches, and finally completely digital switch-

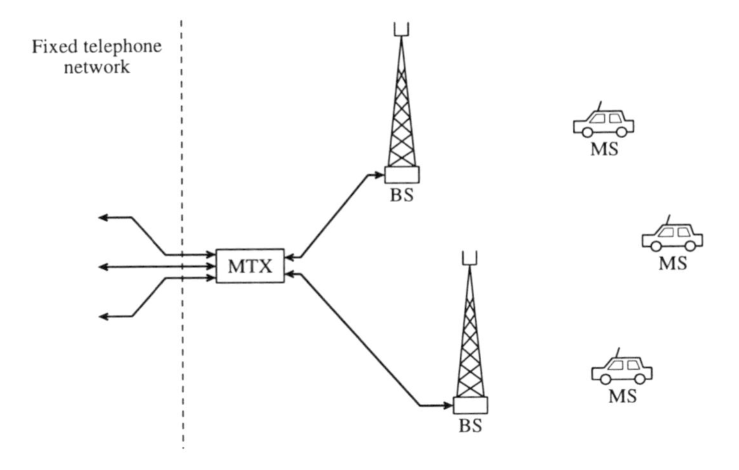

Figure 1.1. Components of a mobile telephone system

ing systems are essential in cellular systems for the speed at which it will switch a call, and for all the functions and features it will offer the user. But while political fights, such as the divestiture of AT&T, slowed things down in the US, European countries continued their own developments.

In Scandinavia, for example, the NMT (Nordic Mobile Telephone) group was formed in 1969, aiming at the development of a new, pan-Nordic mobile telephone system. The way the telephone instrument is used should be as close as possible to the regular telephone. Therefore the roaming concept was introduced in 1973. Roaming makes it possible to keep track of the subscriber anywhere within the Nordic counties. The first network diagram was presented, as shown in figure 1.1.

The mobile station (MS) is installed in cars and trucks, and connects over radio to the nearest base station (BS). These base stations in turn are connected over fixed facilities to a Mobile Telephone Exchange (MTX). From the MTX, access is provided to the public network. An MTX is connected to other MTXs either by direct circuits, or over switched circuits in the public network. These facilities are also used when performing the roaming function. The Scandinavian system operated in the 450-MHz band and was officially inaugurated on October 1st 1981. But the world's first cellular system, be it based on the NMT 450 standard, had been opened for service one month before, in Saudi Arabia. Other countries followed, such as the United States, and all European countries. In the US, the standard was AMPS (American Mobile Phone Standard). The United Kingdom adopted a modified version of this, called TACS (Total Access Communications System). Germany had its Netz C, in Belgium there was MOB-1, etc.

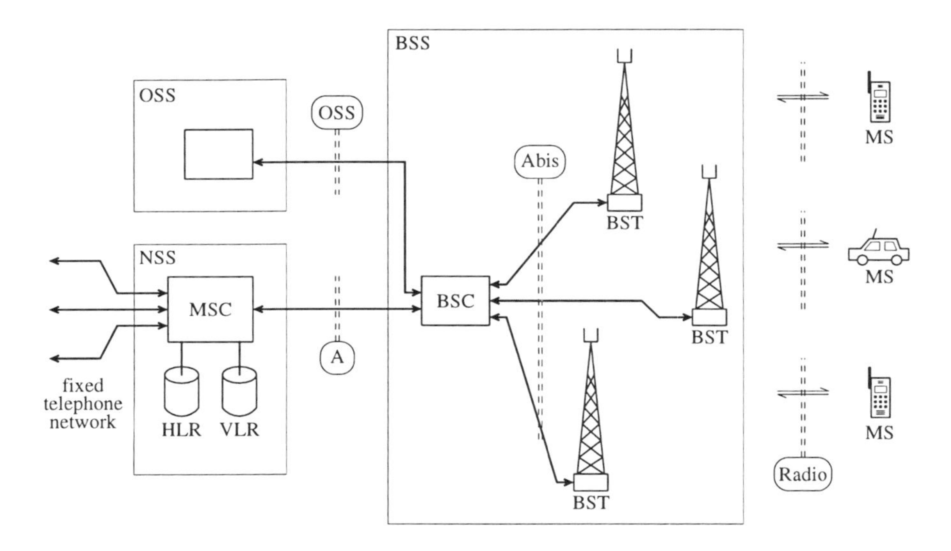

Figure 1.2. Architecture of the GSM system

1.1.2 Digital Cellular Systems

One of the limitations of the above-mentioned systems were the several different standards involved. Roaming between countries with different standards was impossible. Meanwhile, technology was evolving and the increasing number of users created the need for systems with a much more efficient use of the frequency spectrum. As early as 1978, the Conférence Européenne des Postes et Télécommunications (CEPT), that is an organization of European PTTs, decided to reserve two blocks of 25 MHz in the 900-MHz band for mobile communications in Europe. At the CEPT conference in Vienna in 1982, a new standardization working group was created : the Groupe Spécial Mobile (GSM). In 1988, the European Telecommunications Standards Institute (ETSI) was formed, and the GSM standardization activities were transferred into it. ETSI is not restricted to a membership from PTTs, but includes members from user groups and manufacturers. It became clear that in most countries there would be more than one operator : there would be competition. This is reflected in some specific characteristics of the GSM specifications. The GSM system will be build up around open interfaces. This is shown in figure 1.2.

The Mobile Station (MS) includes hand-held or car-borne systems. The Base Station Subsystem (BSS) comprises the Base Station Transceivers (BST) and the Base Station Controller (BSC). The Network and Switching Subsystem (NSS) includes the Mobile Service Switching Center (MSC), and the databases HLR (Home Location

Register) and VLR (Visitor Location Register). These databases are used for roaming. The Operation Sub-System (OSS) supports network management, subscription management, charging and billing, and mobile equipment management. The four principle interfaces are indicated : the Radio interface, the Abis interface, the A interface and the OSS interface. The open interfaces mean that any operator can choose different manufacturers for the different subsystems of his network.

The GSM system operates in two frequency bands around 900 MHz. From 890 to 915 MHz, the MS transmits and the BST receives, whereas from 935 to 960 MHz, the BST transmits and the MS receives. Optimal spectral efficiency is achieved with a narrowband TDMA (Time Division Multiple Access) technology. The channel width is 200 kHz, allowing 124 channels to be placed in the 25-MHz band. Digital speech is transmitted at a rate of 13 kb/s, with a time multiplexing factor of 8. Later, in 1990, a version of GSM was adapted to operate in two 75-MHz wide bands around 1.8 GHz, called DCS-1800 (Digital Communication System 1800). Nowadays, the GSM system is adopted in most European countries, and is also widespread in the Asian region and in Australia. The United States has a digital system called DMPS, and in Japan there is the PHP (Personal Handy Phone) standard.

Of course, this evolution in standards and systems would not have been possible without a parallel evolution in technology. Although battery technology is also improving, putting more energy in smaller volumes with less weight, the main improvements have been made in IC technology. The ever-decreasing line width of CMOS processing technology has gone from around 10 μm in 1970, over 1 μm in the late 1980s, to a present-day value of 0.35 μm in mass production. It seems this will not stop soon, predicting a feature as low as 0.07 μm somewhere around the year 2010 [Geppe Spec96]. This has allowed to implement a lot of the digital baseband signal processing on a single IC. The Radio-Frequency (RF) front-end processing is also realized on integrated circuits, either in GaAs, bipolar, BiCMOS or plain CMOS. All these trends are aiming at two goals : an overall reduction in weight and volume, and a reduction in power consumption which is equivalent to an increase in talk time.

The next section will focus on only a small part of the mobile communication system, i.e. the RF front-end of the mobile terminal. It may be only a small part, but yet it is a very important one since this is where the mass market is. It can therefore count on a lot of interest from industry and other related organizations. Reducing the size, the power consumption or the production cost of this block translates directly into better and/or cheaper products, including higher profits for the manufacturer. We hope to give an introduction in the technology trends currently seen in the integration of the RF front-end signal processing blocks by discussing the several options to be taken in the design of a high-frequency transceiver for a cellular telecommunication system.

1.2 INTEGRATING A TRANSCEIVER

The only way such a rapid growth from mainly professional users to a real mass market product was possible, is the fact that mobile terminals that have a low mass and volume, long talk and standby times, and most of all a low cost became available . These three aspects (low volume, low power and low cost) are only feasible if a high degree of integration is possible. Therefore transceivers are migrating very fast from a three- or four-chip solution with several external components, to a two chip-set solution (one for the analog and one for the digital signal processing) with some external components [Sato ISSCC96, Irie ISSCC97, Heine ISSCC97].

A transceiver (transmitter-receiver) is a building block that interfaces between the user and the transmission medium, i.e. the free air in the mobile communication systems we're talking about. It generally consists of three blocks [Crols 97]. The user-end interfaces between the user information and the digital data representation. The back-end modulates and demodulates the digital data to and from an analog baseband signal that is suited for the transmission technique used (GMSK, QPSK, etc.). The third block is the front-end. This is the building block of interest for us. It mainly does frequency conversions between the high-frequency antenna signals and the low-frequency baseband signal. Its two separate functions are obvious, i.e. receiving and transmitting.

Two well-known receiver topologies exist, i.e. the heterodyne or IF receiver and the homodyne or zero-IF receiver. They are shown in figure 1.3. They both amplify the antenna signal in a Low-Noise Amplifier (LNA), after removing unwanted signals in an RF Band-Pass Filter (BPF). In the IF receiver, the signal is then mixed down with the Local Oscillator (LO) signal to a rather high Intermediate Frequency (IF). This IF frequency must be high such that the mirror frequency of the mixer process lies outside of the passband of the antenna BPF. The required quality of this BPF strongly depends on the IF frequency chosen. The wanted channel is then selected with another high-quality BPF, and digitized in the A/D Converter (ADC) after some Automatic Gain Control (AGC). This receiver architecture is well known and often-used, but it has some major drawbacks when full integration is the goal. These problems are mostly situated in the band-pass filters, as they require rather high quality factors and cannot be integrated without consuming excessive amounts of power. Therefore external BPFs are used, mostly Surface Acoustic Wave (SAW) filters, which increase the cost of the transceiver and also the power consumption because a lot of signals are going on and off the chip. Even more, as this first IF frequency is sometimes too high to be handled by the digital signal processing, one or more extra stages of IF downconversion must be added, again increasing the number of components and the power [Crols 97].

The zero-IF receiver is shown in figure 1.3(b). Here the RF signal is mixed down directly to baseband with an LO signal of the same frequency. The IF is chosen to be

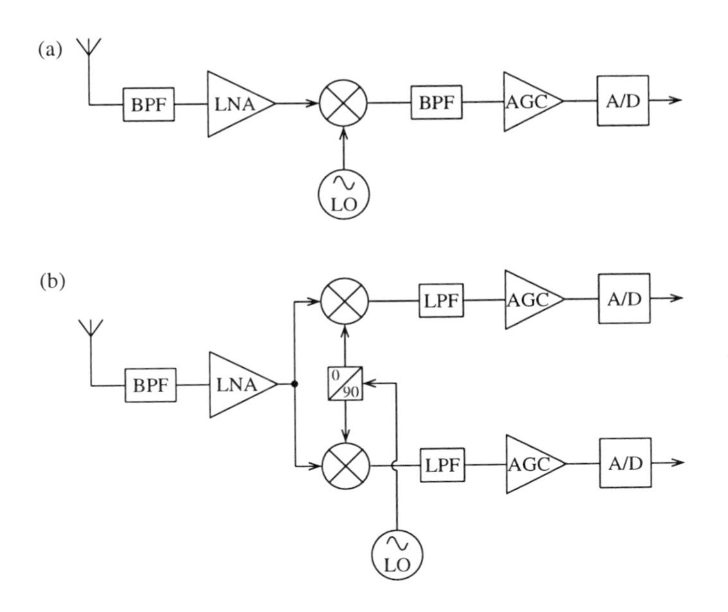

Figure 1.3. Receiver architectures : (a) IF receiver; (b) Zero-IF receiver

zero. Now the mirror signal is the mirror image of the wanted signal itself. To separate them, the downconversion is done in quadrature. The RF signal is mixed twice, once with the LO signal as a sine wave and once with a 90-degrees shifted LO signal, i.e. a cosine wave. This results in two quadrature signals, I and Q, out of which the original modulating signal can be extracted. The channel selection is done with a Low-Pass Filter (LPF). Because of the use of low-pass instead of band-pass filters, the zero-IF receiver is much more suitable for integration than the IF receiver. The RF BPF only needs to block signals from other frequency bands, and does not need a high quality factor. The channel selection filters are low-pass instead of high-Q band-pass, and can easily be integrated. The biggest problem of zero-IF receivers are parasitic baseband signals that are created during downconversion by DC-offsets, imperfect matching between the two mixer paths, LO-to-RF crosstalk in the mixers, etc.

The problems of zero-IF receivers have driven most designers to the IF approach, despite its low integratability. But other architectures are being developed, trying to combine the advantages of both techniques. Two examples are the wideband-IF receiver [Rudel ISSCC97], and the low-IF receiver [Crols 97].

The transmitter designer also has the choice between several architectures, which resemble the receiver architectures. The heterodyne transmitter also uses an IF frequency and a band-pass filter to remove the image signal. Direct upconversion uses an

LO at the same frequency as the carrier frequency. Again, using an IF diminishes the integratability but the zero-IF approach suffers from cross-talk and self modulation. So most transmitters used are of the IF type.

Another key decision in the design of a transceiver is the choice of the technology. Discrete realizations allow the choice of the best technology for every building block. GaAs circuits achieve the best results in critical blocks such as the low-noise amplifier and the power amplifier. Also bipolar transistors achieve at present f_T values of several tens of GHz, and are therefore often used in mixers, phase shifters, etc. But these two technologies block the path for a further generation of an integrated transceiver, i.e. all analog and digital circuits on the same IC. Since the digital signal processing definitely needs CMOS transistors, the use of a BiCMOS technology seems inevitable for this purpose. BiCMOS has the disadvantage that neither the bipolar nor the CMOS transistors are state-of-the-art. So the frequency performance of the analog bipolar part will be worse, as will be the area and power of the digital CMOS part. So recently the interest in CMOS RF circuits has increased dramatically. If one is able to design the high-frequency analog signal processing circuits in a state-of-the-art submicron CMOS technology, a large step will be taken towards the realization of a real single-chip transceiver. So this is a very important aspect of the work presented in this book : the developed circuits will be integrated in a standard CMOS process, without any external components and without any tuning, trimming, or special processing techniques.

In the rest of this chapter, and in all the following chapters, we will concentrate on the building block that is always required in all receiver or transmitter architectures, i.e. the local oscillator generator. As in the high-Q BPFs, the required quality of the synthesized sine wave is generally too high for a monolithic design and external components must be used. But as the transceiver systems evolve towards more integration, the LO frequency synthesizer will have to follow this trend. So new techniques must be investigated that allow the realization of fully integrated Voltage-Controlled Oscillators (VCOs) and other frequency synthesizer components, without a deterioration in the achieved specifications. They are the subject of this work.

1.3 FREQUENCY SYNTHESIZER TYPES

Generally, three common frequency synthesizer types can be distinguished : the table-look-up synthesizer, the direct synthesizer, and the indirect, or phase-locked, synthesizer [Egan 81]. Each of them will be described shortly in the following sections.

1.3.1 The Table-Look-Up Synthesizer

In this type of synthesizer, also called a digital synthesizer, the sinusoidal waveform is created piece by piece by using the digital values of the waveform stored in a memory. A simplified block diagram is shown in figure 1.4. At the clock frequency, a

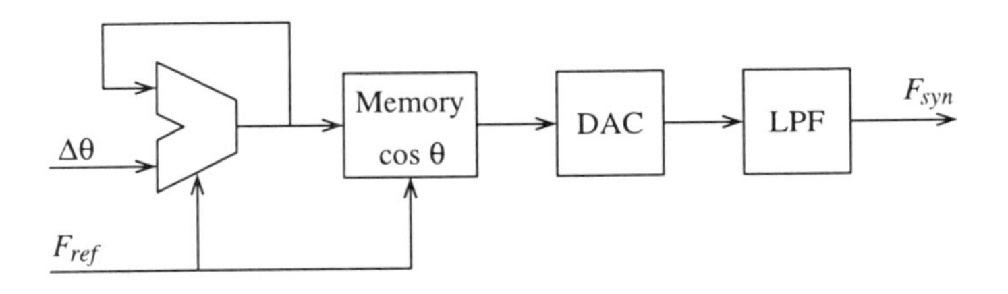

Figure 1.4. Table-look-up frequency synthesizer

number $\Delta\theta$ representing the phase change per clock period is shifted into the accumulator. The capacity of the accumulator corresponds to one complete cycle. The accumulator output thus represents the phase of the signal. This output is used as the address for a memory that contains numbers proportional to the cosines of its addresses. The memory output is converted to an analog voltage using a Digital-to-Analog Converter (DAC), and high-frequency spurs resulting from the digital operation are removed by a Low-Pass Filter (LPF).

For an accumulator with a capacity of N_A bits, the synthesized frequency is

$$F_{syn} = \frac{\Delta\theta}{2^{N_A}} F_{ref}. \tag{1.1}$$

This type of synthesizer permits the phase to be set by setting the number in the accumulator. It can change frequencies very rapidly and fine resolution is easily obtained. However, high frequencies are not possible, due to the limited speed of the memory and high-resolution DACs, and spurs on the output can be a problem.

1.3.2 The Direct Synthesizer

The direct synthesizer employs multiplication, mixing and division to generate the desired frequency from a single reference. An example is shown in figure 1.5. Reference frequencies of 3 and 27-36 MHz are generated from a single crystal reference. Every switch element can provide one of the frequencies 27-36 MHz to the input of the mixers. For example, if the first switch selects the 32-MHz signal ($a = 5$), the input to the second mixer will be $(3 + 32)/10 = 3.5$ MHz. By repeating the mixer-divider block, any desired frequency accuracy is possible [Egan 81].

Of course, this example is not ideal and care has to be taken in choosing the correct reference frequencies, filtering out the unwanted components of the output spectrum, etc. Advantages of this synthesizer type are the possibility for rapid frequency change, and the very pure output spectrum (essentially a replica of the reference oscillator). Disadvantages are the difficult layout of the circuit, in order to limit cross coupling between stages, and large number of components, which causes direct synthesizers to be bulky and of course power-hungry.

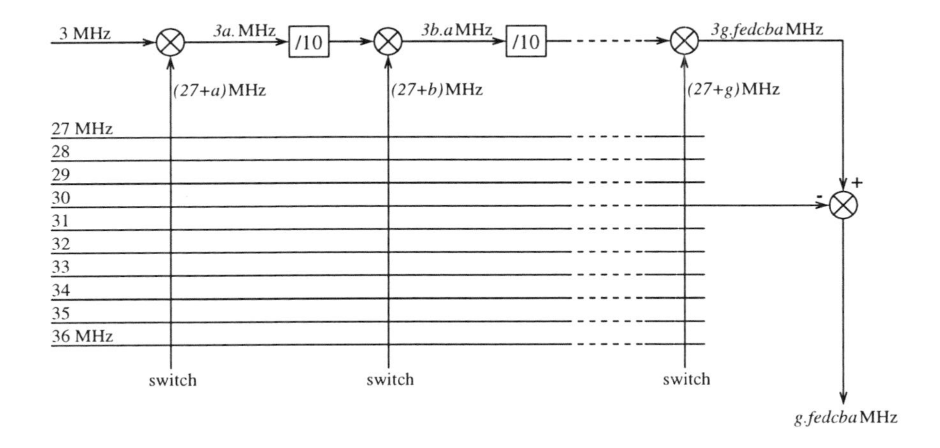

Figure 1.5. Direct synthesizer example

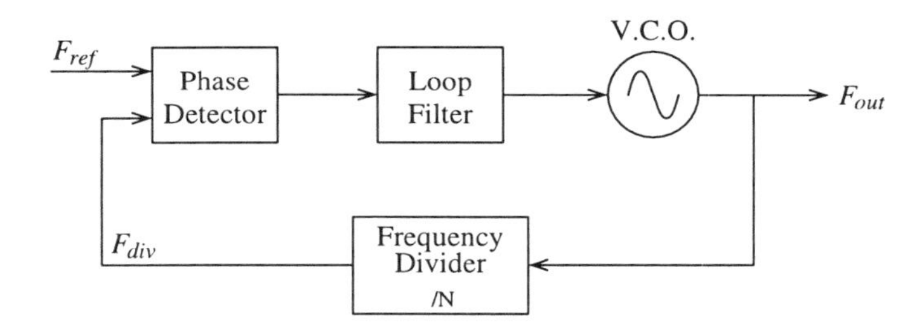

Figure 1.6. General block diagram of a phase-locked loop (PLL) frequency synthesizer

1.3.3 The Phase-Locked Loop Synthesizer

Since the digital frequency synthesizer cannot be used for high output frequencies, and the direct synthesizer is too bulky for silicon integration, all synthesizers used in mobile communication systems are of the phase-locked loop, or indirect, type. In this circuit, the reference frequency is multiplied by a variable number. This is done by dividing the output frequency by that number, and adjusting the output frequency such that this divided frequency is equal to the reference frequency. A simple Phase-Locked Loop (PLL) diagram is shown in figure 1.6.

The Voltage-Controlled Oscillator (VCO) output frequency is divided by a variable number N in the frequency divider. This divided frequency is compared to the reference frequency in the Phase Detector (PD), which gives an output signal equal to the phase difference between its two inputs. The signal is low-pass filtered by the loop filter, and is the control input to the VCO. Under conditions of lock, the two inputs of the phase detector have a constant phase relationship and thus equal frequency. The output frequency therefore is

$$F_{out} = N \cdot F_{ref}. \tag{1.2}$$

If the output frequency increases, the phase difference between F_{div} and F_{ref} will drop and the phase detector output will decrease. The VCO is tuned by this to a lower frequency until the correct output frequency is reached. The loop filter suppresses undesired components in the PD output, and has an important effect on noise, acquisition of lock, response speed, loop stability, etc. The correct mathematical description of the loop operation will be discussed in detail in chapter 2.

The phase-locked synthesizer is inherently slower than both other types. Changing the frequency is done by changing the divider modulus N, which results in a slow change of the VCO control voltage as the loop acquires its steady-state operation. The loop filter and the reference frequency play an important role in this evolution. Fast frequency change is only possible when the loop bandwidth is large. Since the loop bandwidth must normally be limited to one tenth of the reference frequency, and also other specifications such as noise determine the loop characteristics, indirect frequency synthesizers will always respond slower than direct or table-look-up synthesizers.

Care must be taken during design, since low-frequency noise generated by the loop components can modulate the VCO control voltage and therefore corrupt the output spectrum. Also, the output spectrum at frequencies further away from the carrier is completely determined by the VCO spectrum. This requires a high-quality VCO with low noise.

So in the design of a PLL the specifications of every component must be chosen with extreme care. The VCO must be of high quality, the frequency divider must have high speed, the phase detector must be accurate and the loop filter must be such that an optimum compromise is reached in switching speed, frequency settling, output noise, etc. Nevertheless, the PLL frequency synthesizer is the only synthesizer type suitable for integration in a standard Integrated Circuit (IC) process and for low-power operation. The output frequency can be high and is only limited by the maximum frequency obtainable in the VCO and the frequency divider. In the rest of this work, we will only consider the PLL synthesizer.

1.3.4 Combination of Techniques

Another technique has been proposed that also offers the possibilities of integration, high output frequency and fast frequency switching : Direct Digital Frequency Synthesis (DDFS) [Abidi CICC94]. Here, a rather low-frequency signal F_{low} is generated using a table-look-up synthesizer, which is than upconverted to the desired RF frequency with a fixed-frequency oscillator signal F_{IF}. This is shown in figure 1.7. The high-frequency oscillator signal is generated from a lower reference frequency by a

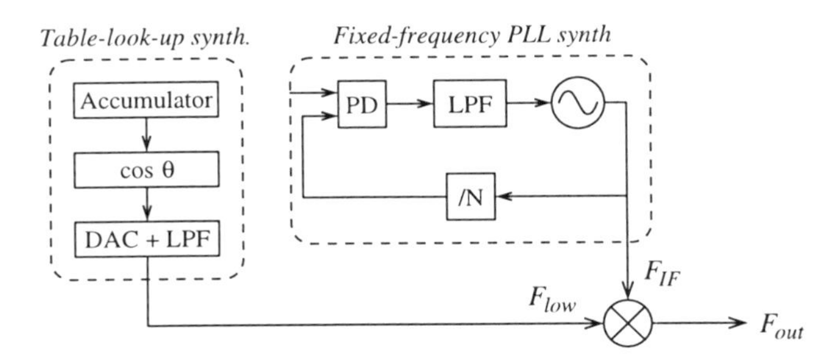

Figure 1.7. Direct digital frequency synthesizer

phase-locked loop. The output frequency is thus

$$F_{out} = F_{low} + F_{IF}. \tag{1.3}$$

The main advantage of this configuration is the fixed frequency of the PLL synthesizer. The loop bandwidth can be optimized for noise, since the reference frequency choice is free, and frequency can be changed rapidly and in small intervals by changing the frequency of F_{low}.

This technique suffers from the limited frequency range due to the low F_{low}, which can only be increased by making a faster DAC. Depending on the required spectral purity of the synthesized signal, the DAC needs a large number of bits and will therefore already be power-hungry. The DDFS will therefore only be useful in systems that require very fast frequency-hopping, such as the one proposed in [Min CICC94].

The applicability of this architecture can be expanded by replacing the table-look-up synthesizer with another PLL. This results in the so-called dual-loop indirect synthesizer, which has more degrees of freedom than the standard PLL architecture. One can choose the reference frequencies and the loop bandwidths of both loops separately, optimizing both of them for the best overall result. The major drawback of this topology is the fact that the two frequencies must be mixed in a single-sideband (SSB) mixer, which requires accurate quadrature phases in both PLLs, low harmonic distortion and well-matched mixers [Razav CICC97].

To reduce the problems of the SSB mixing process, the mixer can be placed inside the loop as shown in figure 1.8. The sidebands resulting from mismatches and harmonics during mixing can be placed such that they are suppressed by the loop filter in the first loop [Razav CICC97].

1.4 THE PRESENTED WORK

As already stated, this work will deal with the problems associated with integrating a phase-locked frequency synthesizer in a standard digital CMOS process. The theory,

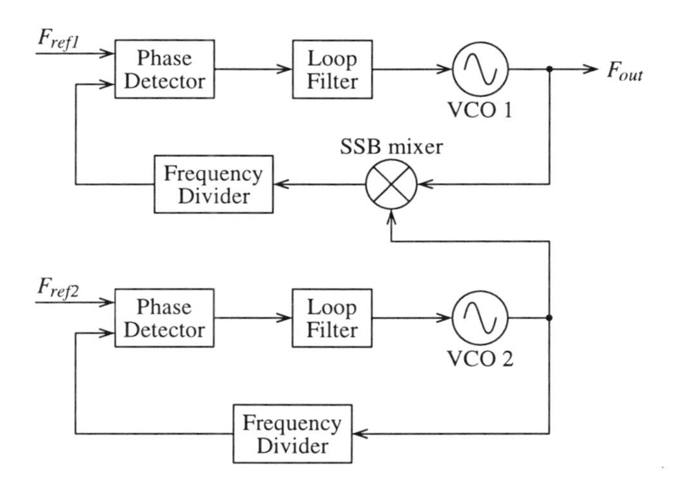

Figure 1.8. Dual-loop PLL frequency synthesizer

mathematical description and operation of a PLL will be discussed in chapter 2. The several building blocks, i.e. the voltage-controlled oscillator, the frequency divider, the phase detector and the loop filter are handled. A lot of attention is paid to the loop filter, which plays an important role in the transient loop characteristic, the output noise spectrum and the loop stability.

In chapter 3, a general theory is developed for calculating the phase noise of oscillators based on the resonance of an inductor (L) / capacitor (C) tank. Indeed, for GSM or DCS-1800 applications, the LO noise requirements can only be achieved with LC-oscillators. Two new concepts are developed, the *effective resistance* and the *effective capacitance*, that allow to evaluate every LC-tank for noise and power consumption. By using several inductors and capacitors, *enhanced LC-tanks* can be designed that offer a trade-off between noise and power.

The two following chapters deal with integrated inductors and their application in an oscillator. Unlike capacitors and resistors, inductors are not readily available in a standard silicon technology. This will have a serious impact on the quality of the LC-tank, which must be as high as possible. Chapter 4 demonstrates the feasibility of *bonding wire-inductance* in a high-quality VCO. Since every chip must be bonded before packaging, bonding wires can be regarded as being a standard feature of IC technology. They offer a low series resistance, and hence low noise. A thorough analysis of the bonding wire inductors is followed by a realization of a state-of-the-art low-phase noise CMOS VCO.

Chapter 5 incorporates two VCO designs that use planar inductors laid out in the standard two levels of metal interconnection available in an IC. This is the most simple way to create integrated inductors, but care must be taken in designing them. Finite-element analysis has allowed to analyze the effects of high-frequency magnetic fields on the losses in the series resistance of the inductor and in the underlying substrate. Based on this, the most optimal inductor for a given specification and frequency range can be designed.

In chapter 6, the other high-frequency building block of a PLL synthesizer, i.e. the frequency divider, is dealt with. Since the VCO frequency is mainly determined by passive elements such as inductors and capacitors, the main frequency limitation of an integrated synthesizer is situated here. Several types of very fast flipflops are investigated, and a new D-flipflop, based on the CMOS equivalent of en ECL flipflop, is developed. Furthermore, a new architecture is proposed that allows the realization of a variable divider modulus without the usual speed degradation with respect to fixed-modulus dividers. Two submicron CMOS designs that use these techniques are presented.

To complete this work, the PLL loop is closed and a completely integrated phase-locked loop frequency synthesizer is presented in chapter 7. The two remaining building blocks, the phase detector and the loop filter, are designed with the aim of creating a DCS-1800 synthesizer. The transient and noise response of the loop are analyzed, and based on this the loop parameters are chosen. Several extra noise sources emerge, all of which are treated to ensure a low-noise output spectrum. A dual-path loop filter is designed that needs smaller capacitor values, which is required to integrate all passive components on chip. A new indirect linearization technique is proposed that introduces no extra phase noise, as opposed to the commonly used direct linearization of the VCO sensitivity.

In a final chapter, some general conclusions drawn from this work are presented.

2 PHASE-LOCKED LOOP FREQUENCY SYNTHESIZERS

2.1 INTRODUCTION

The indirect or phase-locked loop frequency synthesizer offers the distinct advantage over other types of synthesizers of possible low-cost IC integration. This is the main reason why almost all mobile communication chip sets use this circuit. But of course there are also some disadvantages, which require careful study and proper design. Two main areas of specifications characterize a frequency synthesizer.

First, there is the noise characteristic. Ideally, the synthesizer output spectrum should be a Dirac impulse at the desired frequency. But due to several noise influences there will be sidelobes in the frequency spectrum, indicating that there is also a certain amount of signal power at frequencies slightly offset from the desired center frequency. Moreover, due to the operation of some time-discrete components such as the phase detector, several spurious signals can appear in the output spectrum. In a transceiver, the amount of LO power at frequencies away from the wanted carrier must be limited. Otherwise, in the receive mode also strong interfering channels next to the wanted channel will be mixed down. In the transmit mode, one can disturb the signal transmitted by another user in the adjacent channel [Razav JSSC96]. The noise in the output spectrum is determined by the noise of the several PLL building blocks, and by the transfer characteristics of these noise sources to the output, which is mainly ruled by the open loop transfer function.

Secondly, there are the dynamic or transient characteristics of the loop. In a digital cellular system, the mobile terminal must be able to switch very fast between the receive mode in a certain communication channel and the transmit mode at another frequency. The loop must react fast enough to such changes to allow settling of the output frequency with a certain accuracy within a certain time limit. And the steady-state error that remains when all the transients have died away, must be small enough to be within the system's specifications. The loop filter will play a dominant role in this characteristic. Also important are the frequency ranges in which the loop operates correctly. At startup, the loop must be able to lock to a reference frequency that does not correspond to the center frequency of the VCO. Once it is locked, it must be capable of following changes in the reference frequency or the division modulus without loosing the locked status. For these characteristics, the phase detector will turn out to be very important.

To start this chapter, we will present the definition of phase noise in an oscillator signal. The most comprehensible representation discusses noise in the *frequency* domain. However, as the study of a PLL is most easily done using the *phase* of its in- and output signals as loop variables, we will also give the dependencies between these two definitions. In the next section, the general theory is presented which deals with some fundamental characteristics of a PLL. The open loop gain is calculated, and the corresponding noise transfer functions to the output are derived. Section 2.4 gives an overview of the possible implementation options for the four loop building blocks : the phase detector, the loop filter, the voltage-controlled oscillator and the frequency divider. Phase detectors can be analog, such as a multiplier, or sequential, which operate in a more digital way. A lot of attention is paid to the loop filter, which will determine to a great extent the noise and dynamic characteristics of the PLL. The order and the type of the loop are defined by it, and proper design of its poles and zeros allow an optimum open loop gain for the required specifications. Next, some VCO types are discussed as well as the frequency divider implementations.

2.2 DEFINITION OF PHASE NOISE

Phase noise is usually characterized in the frequency domain. For an ideal oscillator operating at frequency ω_0, the output can be expressed as $V_{out}(t) = A \cdot \sin(\omega_0 t + \theta)$, where A is the amplitude and θ is an arbitrary, fixed phase reference. Therefore, its spectrum assumes the shape of a Dirac impulse. In a practical oscillator, however, the output is more generally given by :

$$V_{out}(t) = A(t) \cdot \sin(\omega_0 t + \theta(t)) \quad (2.1)$$

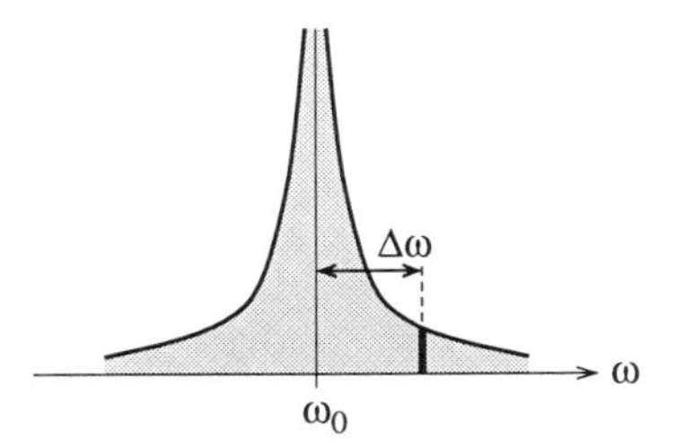

Figure 2.1. Phase noise in an oscillator output spectrum

where $A(t)$ and $\theta(t)$ are now functions of time. As a consequence of the random fluctuations represented by $A(t)$ and $\theta(t)$, the spectrum will have sidebands close to the oscillation frequency. This is shown in figure 2.1.

To quantify this phase noise, we consider a unit bandwidth at an offset $\Delta\omega$ from the carrier, calculate the noise power in the band, and divide this result by the carrier power. This is the single sided spectral noise density in units of deciBell carrier per Hertz $[dBc/Hz]$:

$$\mathcal{L}\{\Delta\omega\} = 10\cdot\log\left(\frac{\text{noise power in a 1-}Hz\text{ bandwidth at frequency }\omega_0 + \Delta\omega}{\text{carrier power}}\right) \quad (2.2)$$

To understand the importance of phase noise in a wireless receiver, consider the situation depicted in figure 2.2. The LO signal used for downconversion has a noisy spectrum such as in figure 2.1, and two transmitters are present, the wanted signal with small power and an unwanted signal in the adjacent channel with a large power level. When these two signals are mixed with the LO output, the downconverted signal will consist of two overlapping spectra. The wanted signal suffers from significant noise due to the tail of the interferer [Razav JSSC96].

In order to get an idea about the phase noise specifications required in a modern cellular phone system, the GSM and DCS-1800 system [ETSI 94] can be taken as an example. In GSM, the minimum signal level that has to be detected is -102 dBm, whereas in the next occupied channel spaced 600 kHz away, an unwanted signal as large as -43 dBm may be present. Assuming a needed SNR after downconversion of 9 dB, the total carrier power in the adjacent channel must be smaller than (-102 dBm) $-$ (-43 dBm) $-$ 9 dB $=$ -68 dBc. Since the channel's bandwidth is 200 kHz, the oscillator's phase noise must be

$$\mathcal{L}_{GSM}\{600\ kHz\} = (\text{-}68\ dBc) - 10\cdot\log(200\ kHz) = \text{-}121\ dBc/Hz \quad (2.3)$$

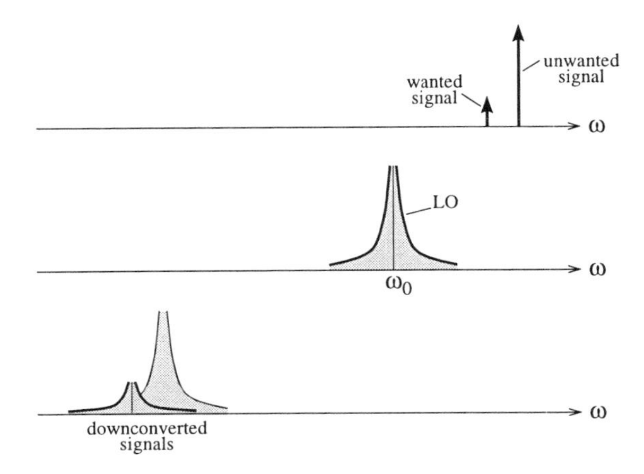

Figure 2.2. Effect of oscillator phase noise in a receiver

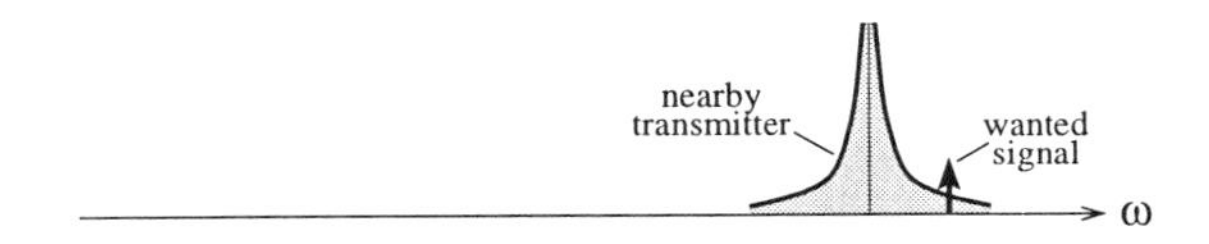

Figure 2.3. Effect of oscillator phase noise in a transmitter

For the DCS-1800 system, the minimum detectable signal level must be -100 dBm, and a similar calculation gives

$$\mathcal{L}_{DCS-1800}\{600\ kHz\} = \text{-}119\ dBc/Hz \tag{2.4}$$

In the transmit path, the effect is slightly different. The situation in figure 2.3 indicates the problem when a noiseless receiver must detect a weak signal at frequency ω_1, while a powerful, nearby transmitter generates a signal at frequency ω_2 with substantial phase noise. The wanted signal will be corrupted by the phase noise tail of this transmitter [Razav JSSC96].

As we will study the general PLL theory using the phase of the signals as the loop variables, it is also clarifying to study the mechanism with which noise in the phase of a sinusoidal signal translates into the frequency spectrum of figure 2.1. Let us consider the effect of a single sinusoidal tone in the phase, $\theta(t) = \theta_p \cdot \sin(\omega_m t)$, where the peak phase deviation, θ_p, is much smaller than one. Equation (2.1) now becomes, assuming

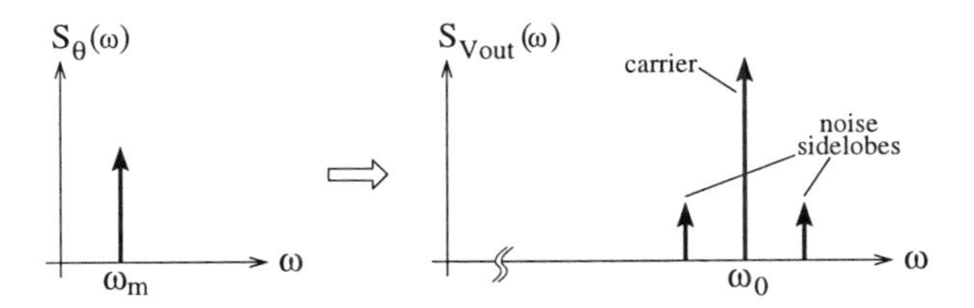

Figure 2.4. Equivalence between phase PSD and single-sided phase noise

a constant amplitude A :

$$V_{out}(t) \approx A \cdot \sin(\omega_0 t) + A \cdot \frac{\theta_p}{2} \cdot \{\sin[(\omega_0 + \omega_m)t] - \sin[(\omega_0 - \omega_m)t]\} \tag{2.5}$$

Therefore the oscillator output voltage power spectral density (PSD) is directly related to the phase noise PSD. Using single sided spectral densities, we have

$$S_\theta(\omega) = \frac{\theta_p^2}{2} \cdot \delta(\omega - \omega_m) \tag{2.6}$$

$$S_{V_{out}}(\omega) = \frac{A^2}{2}\left[\delta(\omega - \omega_0) + \frac{1}{2}S_\theta(\omega - \omega_0) + \frac{1}{2}S_\theta(\omega_0 - \omega)\right] \tag{2.7}$$

This equation shows that the phase noise is directly shifted in frequency towards the carrier and put on the left and the right side of the synthesized frequency. According to (2.2), the phase noise expression is

$$\mathcal{L}\{\omega_m\} = \frac{S_\theta(\omega_m)}{2} \tag{2.8}$$

This process is shown in figure 2.4.

Sometimes it is necessary to relate the PSD of the instantaneous frequency deviations $\Delta\omega(t)$ to the phase noise PSD and to the single-sided phase noise. Since frequency is the derivative of phase, we get

$$S_{\Delta\omega}(\omega) = \omega^2 \cdot S_\theta(\omega) = 2\omega^2 \cdot \mathcal{L}\{\omega\} \tag{2.9}$$

Of course, the noise seen in the frequency spectrum also has its effect in the time domain. Due to the instantaneous frequency deviations, the exact time of one period of the sine wave will differ from period to period. The period has an average value T_p and a timing error $\Delta\tau$. The timing error variance $\sigma_{\Delta\tau} = \sqrt{\overline{\Delta\tau^2}}$ is called *jitter*. A first-order formula to relate jitter to phase noise is [Weiga ISCAS94] :

$$\mathcal{L}\{\Delta\omega\} = \frac{2\pi \cdot \omega_0}{\Delta\omega^2} \cdot \left(\frac{\sigma_{\Delta\tau}}{T_0}\right)^2 \tag{2.10}$$

with T_0 of course the period of the oscillation signal.

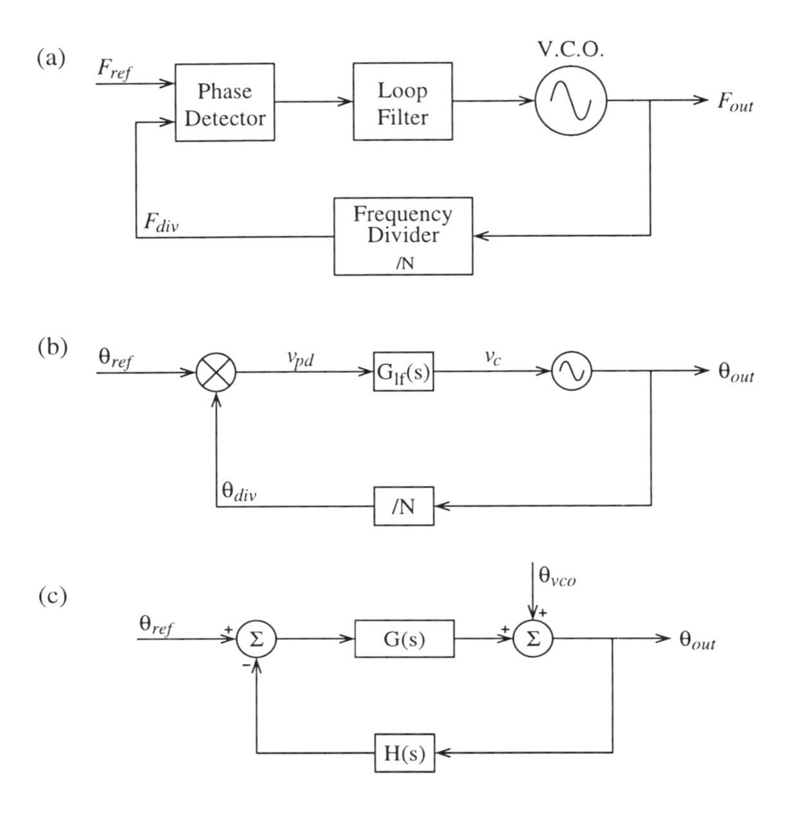

Figure 2.5. PLL frequency synthesizer: (a) Block diagram; (b) State variable diagram; (c) Feedback system

2.3 PLL FUNDAMENTALS

The block diagram of a PLL has already been shown in figure 1.6, and is repeated here in figure 2.5(a). The circuit is called a *Phase*-Locked Loop because the feedback operation in the loop automatically adjusts the phase of the locally generated signal F_{out} to the phase of an incoming signal F_{ref}. As already stated, the easiest way to study this circuit is by using the phase of the reference and the phase of the VCO signal as loop variables. This is indicated in figure 2.5(b).

The input signal has a phase $\theta_{ref}(t)$ and the VCO output has a phase $\theta_{out}(t)$. The prescaler divides the VCO frequency (and hence also the VCO phase) by a factor N :

$$\theta_{div}(t) = \frac{\theta_{out}(t)}{N} \tag{2.11}$$

We assume that the loop is locked and that the phase detector (PD) gives an output voltage proportional to the difference in phase between its inputs :

$$v_{pd}(t) = K_{pd} \cdot [\theta_{ref}(t) - \theta_{div}(t)] \tag{2.12}$$

where K_{pd} is called the phase detector gain factor and is measured in units of V/rad.

This phase error voltage v_{pd} is filtered by the loop filter with its transfer function $G_{lf}(s)$, which normally has a low-pass characteristic. Noise and high-frequency signal components are suppressed. This filter determines to a great extent the noise and dynamic performance of the loop.

The VCO frequency is determined by the control voltage v_c. The deviation of the VCO frequency from its center frequency is $\Delta\omega = K_{vco} \cdot v_c$ where K_{vco} is the VCO gain factor in units of $[rad/Vs]$. Since frequency is the derivative of phase, the VCO operation can also be described as

$$\frac{d\theta_{out}(t)}{dt} = K_{vco} \cdot v_c(t) \tag{2.13}$$

By taking the Laplace transforms we obtain

$$\theta_{out}(s) = \frac{K_{vco} \cdot V_c(s)}{s} \tag{2.14}$$

In figure 2.5(c) the PLL is depicted as a standard feedback network with forward transfer function $G(s)$ and feedback factor $H(s)$. The open loop transfer function equals

$$GH(s) = \frac{K_{pd} \cdot G_{lf}(s) \cdot K_{vco}}{N \cdot s} \tag{2.15}$$

This transfer function can be used in the study of the loop response to noise and transients.

2.3.1 Noise Characteristics

Basically, the noise in the PLL output signal will largely be determined by two noise sources : noise coming from the reference signal, and noise generated in the VCO itself, which is represented by the input signal $\theta_{vco}(s)$ in figure 2.5(c). We can calculate the response of the output signal to both of these sources. The closed-loop response to a VCO noise signal is

$$\frac{\theta_{out}(s)}{\theta_{vco}(s)} = \frac{1}{1 + GH(s)} = \frac{N \cdot s}{N \cdot s + K_{pd} \cdot G_{lf}(s) \cdot K_{vco}} \tag{2.16}$$

whereas the closed-loop response to a reference noise signal is

$$\frac{\theta_{out}(s)}{\theta_{ref}(s)} = \frac{G(s)}{1 + GH(s)} = \frac{N \cdot K_{pd} \cdot G_{lf}(s) \cdot K_{vco}}{N \cdot s + K_{pd} \cdot G_{lf}(s) \cdot K_{vco}} \tag{2.17}$$

Suppose initially that the loop filter has a constant transfer function, i.e. $G_{lf}(s) = K_{lf}$. The open loop transfer function than becomes

$$GH(s) = \frac{K_{pd} \cdot K_{lf} \cdot K_{vco}}{N \cdot s} = \frac{K_F}{N \cdot s} \tag{2.18}$$

K_F is the forward gain of the PLL and has units of s^{-1}.

Equations (2.16) and (2.17) then reduce to

$$\frac{\theta_{out}(s)}{\theta_{vco}(s)} = \frac{1}{1 + K_F/(Ns)} = \frac{s}{s + \omega_c} \tag{2.19}$$

$$\frac{\theta_{out}(s)}{\theta_{ref}(s)} = \frac{K_F/N}{1 + K_F/(Ns)} = N\frac{\omega_c}{s + \omega_c} \tag{2.20}$$

where ω_c is defined as the cross-over frequency, i.e. the frequency at which the open loop gain has a magnitude equal to one :

$$\omega_c = \frac{K_F}{N} \tag{2.21}$$

So the noise transfer function from the VCO to the output has a high-pass characteristic. Noise at high frequencies passes unattenuated, because the feedback action of the loop is too slow to suppress these noise components. At lower frequencies there is a first-order roll-off, as the loop feedback becomes stronger with smaller frequencies. The 3-dB cut-off frequency of this characteristic is ω_c. This situation is depicted in figure 2.6(a). The solid line represents the typical output noise power spectral density (PSD) of an oscillator. This graph will be studied in great detail in chapter 3, but until then we can work with a typical graph like this. Three regions can be distinguished. At high frequencies, a flat noise floor is obtained. The most important region is the one where the phase noise decreases quadratically with the offset frequency (ω^{-2}). This region originates from white noise sources around the carrier frequency that are amplified to a certain level due to the positive feedback in the oscillator. Finally, a ω^{-3} region exists, which is the result of low-frequency $1/f$ noise that is upconverted to the carrier frequency by non-linearities in the amplifier. The dotted line represents the output noise PSD, which indeed follows (2.19).

Noise from the reference has a low-pass characteristic with the same 3-dB cut-off frequency ω_c superimposed on a multiplication with factor N. The resulting PSDs are depicted in figure 2.6(b). The reference noise has the same shape of curve as the VCO noise, but due to the high-quality source used, the magnitude is generally lower. This is annihilated by the fact that the noise is multiplied by N for frequencies lower than ω_c, as can be seen in the dotted line of the output noise PSD. To achieve an optimal noise performance, both graphs in figure 2.6 must be studied, and the loop parameters must be optimized carefully to minimize the total output noise.

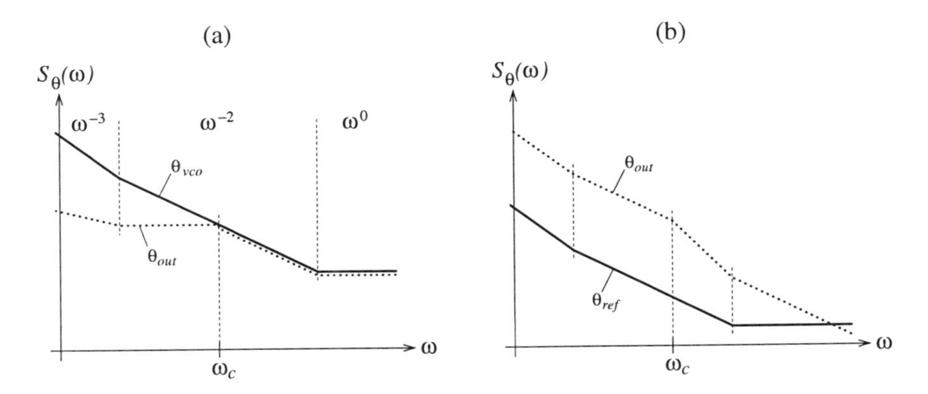

Figure 2.6. Phase noise transfer functions in a PLL synthesizer : (a) VCO noise; (b) Reference noise

We call this simplest PLL (with a constant loop filter) a type-1 loop because the open-loop gain has only one pole at zero frequency. It is also a first-order loop because there is only one significant pole. If we add a first-order low-pass filter $G_{lf}(s)$, the loop will become a second-order (total number of poles), type-1 (number of poles at zero frequency) loop. Since the VCO always introduces a pole at $s = 0$, every PLL is of at least type-1. Extra poles are only introduced in the loop filter, so this will determine the type and the order of the PLL.

2.3.2 Transient Characteristics

The derivations of the previous section were based on the steady-state operation of the PLL : a constant reference frequency is applied and the division modulus is fixed. However, the design of the PLL also incorporates the dynamic behavior of the loop at startup or when the reference frequency and/or the division modulus is changed. This will be discussed in this section.

2.3.2.1 Tracking. First, we will study the phase error $\theta_{err}(t) = \theta_{ref}(t) - \theta_{div}(t)$ that results from a specified input $\theta_{ref}(t)$. We will assume that the loop is in lock, and that the phase error is sufficiently small to justify an assumption of linearity.

We are most interested in the steady-state phase error $\theta_{err,ss}$, i.e. the value that $\theta_{err}(t)$ assumes after all the transients have died away. It can be readily evaluated by means of the final value theorem of the Laplace transformation which states that

$$\lim_{t\to\infty} \theta(t) = \lim_{s\to 0} s\theta(s) \tag{2.22}$$

So if the reference input has a phase step of magnitude $\Delta\theta$, the resulting steady-state phase error for a first-order loop is

$$\theta_{err,ss} = \lim_{s\to 0}\left[s\cdot\theta_{ref}(s)\cdot\frac{1}{1+GH(s)}\right] = \lim_{s\to 0}\left[s\cdot\frac{\Delta\theta}{s}\cdot\frac{1}{1+K_F/(Ns)}\right] = 0 \tag{2.23}$$

So the PLL will reduce any phase error to zero. If the input frequency changes with a step of size $\Delta\omega$, the input phase equals $\theta_{ref}(t) = \Delta\omega\cdot t$. The resulting steady-state phase error is

$$\theta_{err,ss} = \lim_{s\to 0}\left[s\cdot\frac{\Delta\omega}{s^2}\cdot\frac{1}{1+K_F/(Ns)}\right] = \frac{\Delta\omega\cdot N}{K_F} = \frac{\Delta\omega}{\omega_c} \tag{2.24}$$

This is the situation that appears when suddenly the division modulus in the synthesizer is changed in order to obtain a new output frequency. In order to keep the final error small, the loop filter gain must be as high as possible.

Besides the steady-state behavior, it is also necessary to determine the transient phase error caused by particular inputs. Again, the most important situation in a frequency synthesizer is the one where the division modulus N is changed, which is equivalent to a change in input frequency $\Delta\omega$. How long does it take for the output frequency to reach the new desired value ? For a first-order loop the following calculations can be made :

$$\theta_{err}(s) = \frac{1}{1+GH(s)}\cdot\theta_{ref}(s) = \frac{s}{s+\omega_c}\cdot\frac{\Delta\omega}{s^2} = \frac{\Delta\omega}{s\cdot(s+\omega_c)} \tag{2.25}$$

$$\theta_{err}(t) = \frac{\Delta\omega}{\omega_c}\cdot\left(1-e^{-\omega_c t}\right) \tag{2.26}$$

So the final frequency is obtained after an exponential behavior with time constant $\tau_c = 1/\omega_c$. With these equations, one can readily calculate the time T_ε it takes to settle the loop to the new output frequency within a specified accuracy ε (e.g. 0.1%). It is given by

$$T_\varepsilon = -\frac{\ln\varepsilon}{\omega_c} \tag{2.27}$$

In a system such as GSM or DCS-1800, the LO has to switch from the receive channel to the transmit channel. The frequency step magnitude is in the order of 100 MHz, and settling must be done within 100 Hz ($\varepsilon = 10^{-6}$). With a 1-kHz loop bandwidth, the settling time equals 2.2 $msec$.

2.3.2.2 Acquisition. In the previous derivations, it is assumed that the loop is already in lock. But every loop starts in an unlocked condition at power-up, and must be able to achieve lock, either by its own natural actions or with the help of auxiliary circuits. This process is called acquisition. Since acquisition is inherently a non-linear process, analytical calculation goes beyond the scope of this work. We will limit ourselves to some descriptive analysis. More information can be found in [Egan 81, Gardn 79].

If the initial VCO frequency is close enough to $N \cdot F_{ref}$, the PLL will lock up with just a phase transient. No cycles will be missed before lock is obtained. The frequency range over which this is possible is called the *lock-in range* $\Delta\omega_L$.

The *pull-in range* $\Delta\omega_P$ is defined as the range of frequencies over which the PLL will acquire lock after slipping cycles for a while. It is thus always larger than the lock-in range.

Finally, the frequency range over which the loop will hold lock is defined as the *hold-in range*. It is larger than both other defined ranges. For example, in (2.24) the linear approximation of phase error due to a frequency offset is shown to be $\theta_{err,ss} = \Delta\omega/K_F$. However, a real phase detector does not have an infinite linear range, as will be discussed in section 2.4.1. For a sinusoidal-characteristic PD the true expression should be

$$\sin\theta_{err,ss} = \frac{\Delta\omega}{K_F} \tag{2.28}$$

Since the sine function cannot exceed unit magnitude, there is no solution for $\Delta\omega > K_F$. The hold-in range therefore equals $\Delta\omega_H = \pm K_F$. Other types of phase detectors have a larger linear range and can therefore extent the hold-in range. Of course, these definitions are only valid as long as the limit is set by the PD and not by some other nonlinearity, such as clipping in an operational amplifier or of the VCO control signal.

2.4 PLL BUILDING BLOCKS

This section will describe shortly the several implementations that exist for the PLL components. They include the phase detector, the loop filter, the oscillator and the frequency divider. More information can be found in [Egan 81, Gardn 79, Meyr 90].

2.4.1 Phase Detector

Three categories of phase detectors can be distinguished. Analog PDs or multipliers rely on the DC component that results when multiplying two sinusoidal waveforms of the same frequency. Sequential circuits, such as the EXOR and the flip-flop PD, operate on the information contained in the zero crossings of the input signals. The third category, the phase-frequency detector, is actually also a sequential circuit, but also provides a frequency-sensitive signal to aid acquisition when the loop is out of lock.

2.4.1.1 Multipliers. If both inputs to the PD are sinusoidal, a mixer or multiplier can be used. Two input signals $A_1 \sin(\omega_1 t + \theta_1)$ and $A_2 \cos(\omega_2 t + \theta_2)$ result in multiplier output signal

$$v_d = A_d \cdot \{\sin[(\omega_1 - \omega_2)t + \theta_1 - \theta_2] + \sin[(\omega_1 + \omega_2)t + \theta_1 + \theta_2]\} \quad (2.29)$$

At phase lock, both frequencies are the same and the DC component of the PD output equals $A_d \sin(\theta_1 - \theta_2)$. This is proportional to the phase difference at small values of $(\theta_1 - \theta_2)$ and is the useful signal component. A lot of undesirable components are also present, which must be attenuated by the loop filter. The most obvious is the sum-frequency output, which is situated at twice the reference frequency. Also the reference frequency itself is present in the output, with a magnitude dependent on the LO-to-IF and RF-to-IF isolation of the mixer.

The multiplier PD is especially useful in applications where the reference frequency is too high for other circuits, and where the loop bandwidth is sufficiently narrow, so that the filtering of the undesired components can be effective.

2.4.1.2 Exclusive OR Gate. An exclusive-OR gate can be used as a PD as shown in figure 2.7. The output waveform for two input signals A and B is shown in figure 2.7(b) and has an average value that is proportional to the phase difference over a range of half a cycle, as indicated in figure 2.7(c). The output C does not contain any energy at the reference frequency, but the second harmonic reaches a maximum amplitude of 1.27 times the peak-to-peak range of the PD characteristic at a phase difference of 90°. Unfortunately, this is exactly the operating point that will be chosen as the center of operation, as it is situated at the middle of the linear range.

2.4.1.3 Flip-Flop. The operation of a flip-flop as a phase detector is shown in figure 2.8. Narrow pulses at both inputs A and B set and reset the output C. The average value of C has the shape of a sawtooth, with a linear range of a full cycle. In the middle of this linear range, the output has a component at the reference frequency with a magnitude of 1.27 times the peak-to-peak range of the PD. The harmonic frequency content of the EXOR and the flipflop PD can be compared in figure 2.9. The EXOR PD has an important spurious output signal at twice the reference frequency, whereas the most important harmonic of the flipflop PD is situated at the fundamental of the reference frequency. An advantage of the flipflop PD is its linear range of a full cycle.

2.4.1.4 Phase-Frequency Detector. The Phase-Frequency Detector (PFD) is also a sequential PD, but contains a memory function that allows it to give some information about the frequency as well when the loop is not in lock. It is usually implemented together with a charge pump, as illustrated in figure 2.10(a). The PFD has two outputs, Up and Dn, which open or close the two current sources of the charge pump. An active signal on Up causes the upper current source to be activated, which results

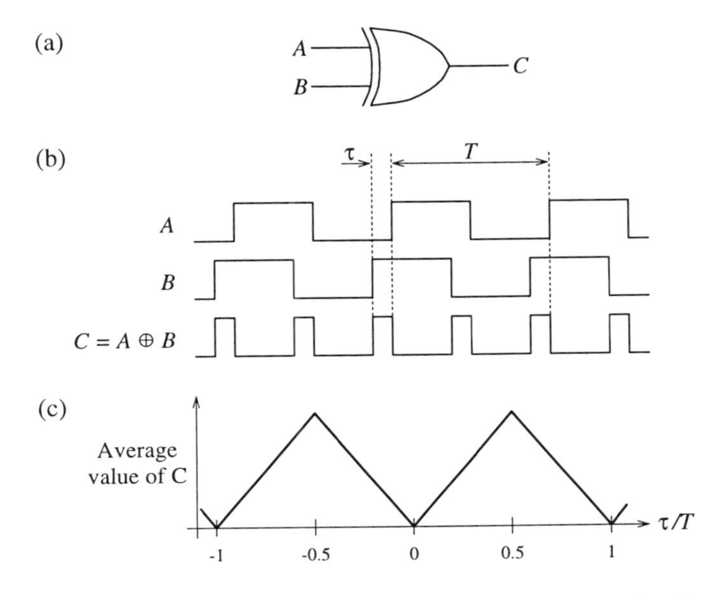

Figure 2.7. (a) EXOR phase detector; (b) Operation; (c) Transfer characteristic

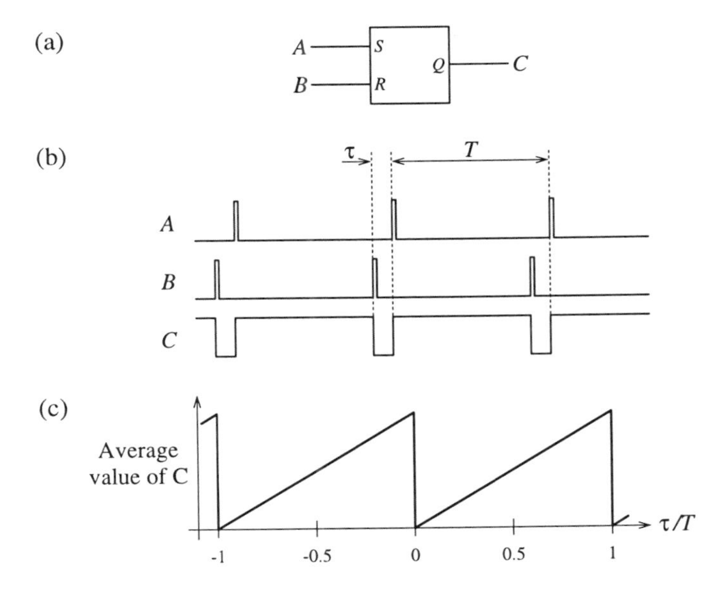

Figure 2.8. (a) Flip-flop phase detector; (b) Operation; (c) Transfer characteristic

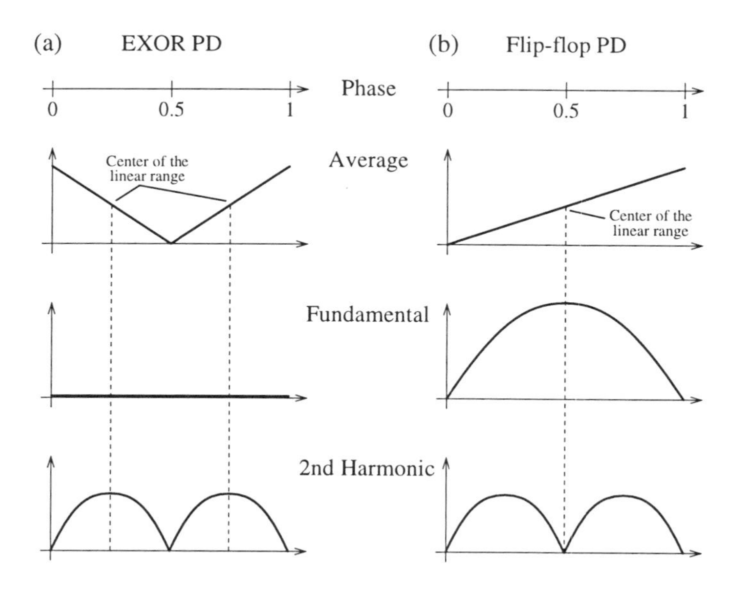

Figure 2.9. Harmonic frequency contents of (a) EXOR PD; (b) Flip-flop PD

in a positive output current I_C. This current causes the output voltage C to rise. This state is a *pump-up* state. An active signal on Dn activates the lower current source, the output current I_C is negative, and the voltage C drops. This is the *pump-down* state. The third possible charge pump state is the one with none of the signals Up or Dn active. The output current is zero and node C is a high-impedance node. The fourth state, i.e. the one with both current sources active, never occurs theoretically. The output current is transferred to a voltage in the impedance Z_{lf}.

The operation principle of the PFD is shown in figure 2.10(b). The reference pulse causes the output to change in a positive direction, unless the output is already positive, in which case the pulse has no effect. Similarly, the loop's divider output causes a negative transition unless the output is already negative. The average voltage output versus phase is plotted in figure 2.10(c). The linear phase range is 720°. At lock, the output contains no spurious signals at all, because the up- nor the down-pulse occurs.

The most important problem of the PFD is the crossover distortion, changes in gain that occur near zero phase error [Egan 81, Hill RFD92]. If both the reference and the divider pulse appear at the same time, none of the outputs becomes active and the charge-pump output is in a high-impedance state. Even if the phase difference changes slightly, the PD will not immediately respond to this since it requires some finite time for the Up and Dn pulses to propagate through the circuit. So the charge pump keeps its high impedance state although there is a slight phase difference. The PD charac-

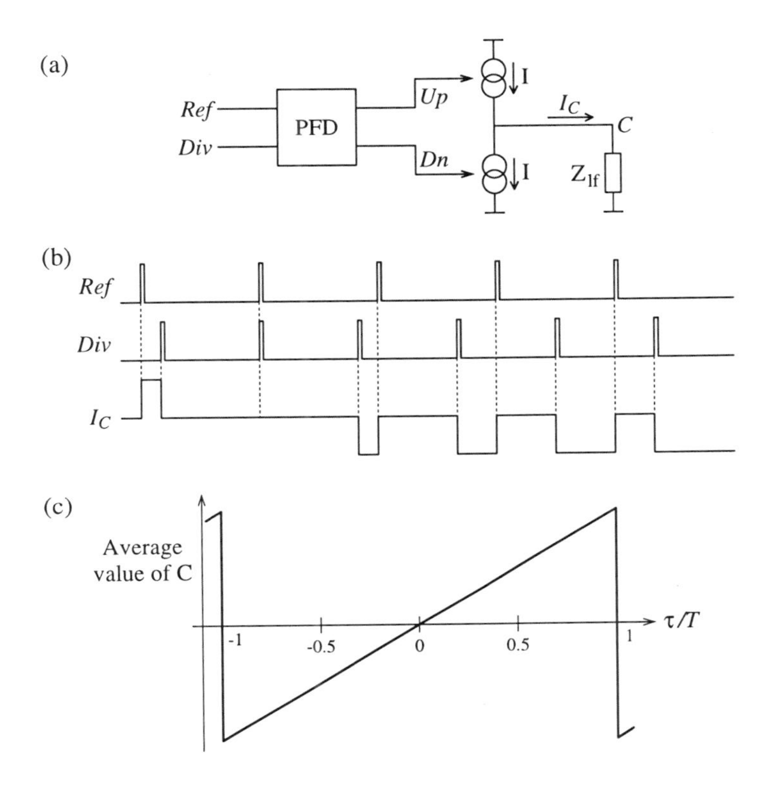

Figure 2.10. (a) Phase-frequency detector; (b) Operation; (c) Transfer characteristic

teristic actually has a flat response, a "dead zone", near zero phase difference. The PLL loop is effectively opened (since K_{pd} is zero) and the output spectrum changes accordingly.

This phenomenon is remedied by giving a fixed minimum width to both the pump pulses. An example of a PFD circuit that performs this function is shown in figure 2.11 [Gardn 79, Mijus JSSC94]. The output terminals Up and Dn are active low. The delay block after the 4-input NAND determines the minimum width of the up- and down-pulse. If the divider output lags the reference signal, the up-signal will become active for a certain time. This time equals the time difference between the two signals, plus the delay through the circuit, including the one of the extra added delay block. Due to this non-zero delay in the circuit, also the down-pulse will become active for a short time, equal to the delay of the signal through the circuit and the delay block. The net difference between up- and down-time determines the total change in charge pump output voltage, and is proportional to the phase difference between the two PFD

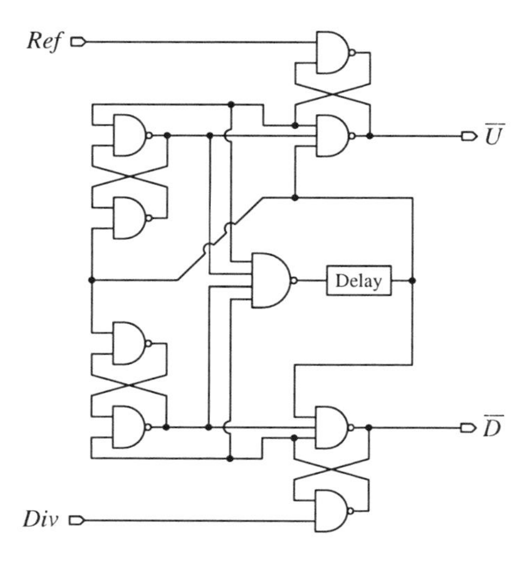

Figure 2.11. Circuit implementation of a phase-frequency detector without dead zone

inputs. The same holds for a divider signal that leads the reference signal. Even if there is no phase difference between the two inputs, both the up- and down-signals are active during a short period determined by the delay-block in the schematic. As the charge injected into the impedance Z_{lf} by these two pulses is of equal magnitude and opposite phase, the net effect is zero, as it should be for zero phase difference. But very important, the charge pump does not stay in its high-impedance state and the loop is always closed.

2.4.2 Loop Filter

Most of the PLL's specifications will be determined by the loop filter. In the loop filter, extra poles and zeros can be introduced in the open loop transfer function, which are used to set the noise and transient performance of the PLL.

2.4.2.1 First-Order PLL. First, we will repeat the case that was already mentioned in section 2.3, where the loop filter has a constant gain K_{lf} over the full frequency range. The open loop gain is given by (2.18) and its Bode plot is shown in figure 2.12. There is one pole at zero frequency, so the magnitude drops with 20 dB/dec and the phase is constant at -90°.

The noise characteristic of the PLL output signal is determined primarily by the response to noise coming from the reference and from the VCO. The noise transfer functions are given by equations (2.19) and (2.20).

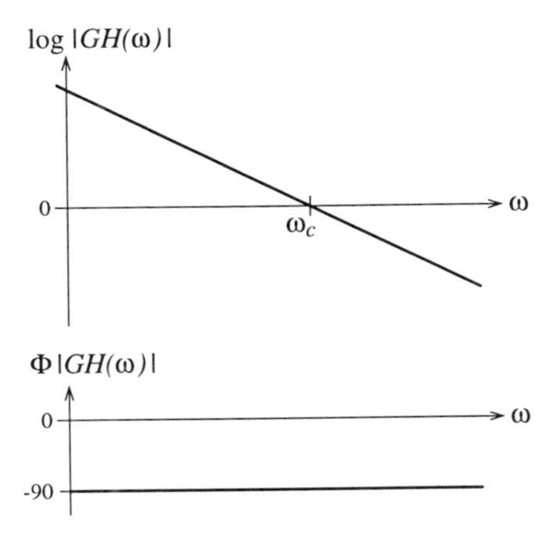

Figure 2.12. Open loop gain Bode plot for a 1-st order, type-1 PLL

The dynamic behavior can be investigated by calculating the reaction of the PLL to a change in input phase or input frequency. The resulting steady-state phase error is given by equations (2.23) and (2.24).

2.4.2.2 Second-Order PLL. In order to stop phase detector signal components at the reference frequency, an extra pole is usually added at a frequency larger than ω_c. This does not change the general noise performance or the steady-state phase error significantly, but reduces spurs in the output signal at the reference frequency and its harmonics. Because the reference frequency is usually not very high compared to the loop bandwidth, the filter pole must be low enough to have an appreciable effect on the loop response. The PLL is now a second-order, type-1 loop. The open loop gain Bode plot is drawn in figure 2.13.

With a loop filter transfer function equal to

$$G_{lf}(s) = \frac{K_{lf}}{1 + s/\omega_p} \tag{2.30}$$

the closed-loop transfer function from θ_{ref} is

$$\frac{\theta_{out}(s)}{\theta_{ref}(s)} = \frac{G(s)}{1 + GH(s)} = \frac{\omega_p K_F}{s^2 + \omega_p s + \omega_p K_F / N} \tag{2.31}$$

In terms of standard notation, this becomes

$$\frac{\theta_{out}(s)}{\theta_{ref}(s)} = \frac{\omega_p K_F}{s^2 + 2\zeta \omega_n s + \omega_n^2} \tag{2.32}$$

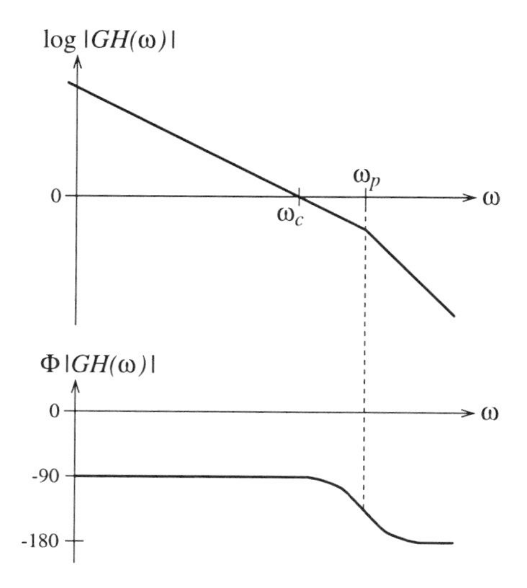

Figure 2.13. Open loop gain Bode plot for a 2-nd order, type-1 PLL

with a natural frequency of

$$\omega_n = \sqrt{\frac{\omega_p K_F}{N}} = \sqrt{\omega_c \omega_p} \tag{2.33}$$

and a damping factor equal to

$$\zeta = \frac{1}{2}\sqrt{\frac{\omega_p N}{K_F}} = \frac{1}{2}\sqrt{\frac{\omega_p}{\omega_c}} \tag{2.34}$$

This mathematical derivation leads to what is generally known for second-order systems : in order for the loop to be relatively stable, the phase margin of the system must be high enough, i.e. the second pole must be a certain factor higher than the loop bandwidth. If ω_p becomes too low, the transient response of the loop becomes less damped, which causes ringing.

We can also conclude that a second-order PLL can never become unstable, since two poles can never cause a phase shift larger than 180° for any finite frequency. However, because of several parasitics, the two-pole system is only an approximation.

2.4.2.3 Third-Order PLL. In order to improve the transient characteristics of the PLL, a low-frequency pole is often introduced in the loop filter [Egan 81]. Since this gives an extra phase shift of 90°, a compensating zero must be introduced in order to

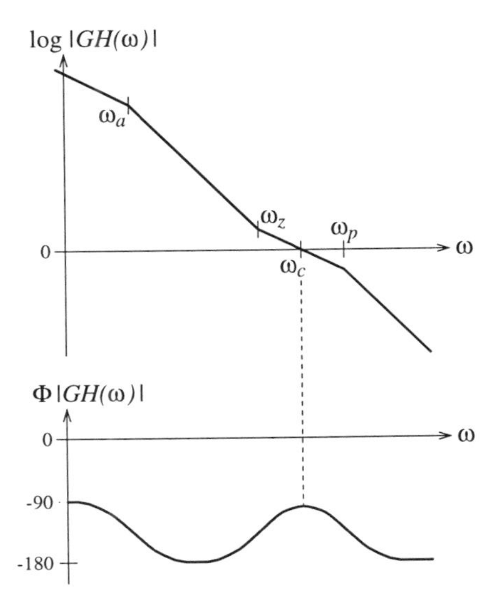

Figure 2.14. Open loop gain Bode plot for a 3-rd order, type-1 PLL

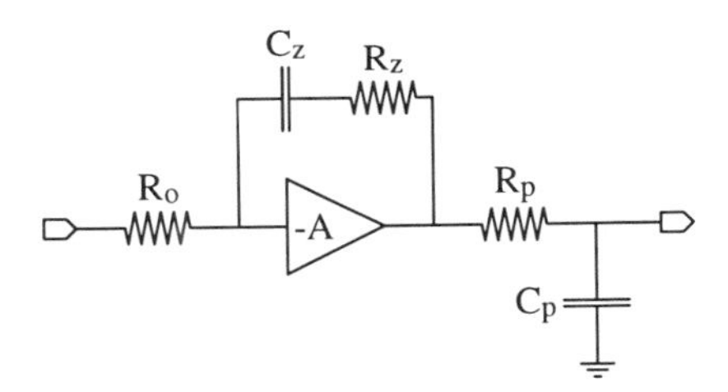

Figure 2.15. Loop filter implementation for a 3-rd order, type-2 PLL

keep the phase margin high enough. The Bode plot for this third-order, type-1 PLL is shown in figure 2.14.

Again, to ensure sufficient stability, the compensating zero must be placed low enough to guarantee a phase margin larger than e.g. 60°. Because of the higher gain at low frequencies, this loop can better track a change in input frequency [Meyr 90]. Ideally, the low-frequency pole can be moved to zero, which results in a type-2 PLL. In this case, the steady-state phase error in response to a frequency change is zero. An active filter like the one shown in figure 2.15 can be used. The pole and zero positions of this filter are given by $\omega_z = (R_z C_z)^{-1}$ and $\omega_p = (R_p C_p)^{-1}$.

2.4.2.4 Charge-Pump PLL. Phase-locked loops incorporating a phase-frequency detector with a charge pump as shown in section 2.4.1.4 are widely used. The reasons for their popularity are their availability as low-cost ICs [MC4044], a tracking range of two cycles, and a frequency sensitive error signal. If the pump currents of the charge pump are both equal to I_{qp}, the total transfer function of phase detector and loop filter is

$$K_{pd} \cdot G_{lf}(s) = \frac{I_{qp}}{2\pi} \cdot Z_{lf}(s) \tag{2.35}$$

If the impedance $Z_{lf}(s)$ is the series connection of a capacitor C_z and a resistor R_z, Z_{lf} equals $R_z + 1/sC_z$ and the open loop transfer function becomes

$$GH(s) = \frac{I_{qp}}{2\pi} \cdot Z_{lf}(s) \cdot \frac{K_{vco}}{sN} = \frac{I_{qp} \cdot K_{vco}}{2\pi \cdot N} \cdot \frac{1 + sR_zC_z}{s^2C_z} \tag{2.36}$$

This is the transfer function for a second-order, type-2 PLL. There are two poles at zero frequency, one is caused by the VCO and the other one originates from the high-impedance state of the charge pump when locked. The crossover frequency is

$$\omega_c = \frac{I_{qp} \cdot K_{vco} \cdot R_z}{2\pi \cdot N} \tag{2.37}$$

and the compensating zero is located at

$$\omega_z = \frac{1}{R_zC_z} \tag{2.38}$$

A charge-pump PLL is the only one that needs only a passive filter to make a type-2 loop. With a conventional phase detector, an active filter must be used to obtain a transfer function like this.

At each cycle of the PFD, a pump current I_{qp} is driven into the filter impedance resulting in an instantaneous voltage jump of $I_{qp}R_z$. The corresponding frequency jump is

$$|\Delta\omega| = K_{vco} \cdot I_{qp} \cdot R_z = 2\pi \cdot \omega_c \tag{2.39}$$

which is generally larger than the average frequency increment per cycle [Meyr 90]. However, this effect is not as severe as it seems, because it disappears when the loop is in lock and the charge pump remains in its high-impedance state. But if these spurs cannot be tolerated, a capacitor C_p can be added in parallel to the loop filter impedance as in figure 2.16, resulting in a third-order, type-2 PLL. The impedance of this filter is

$$Z_{lf}(s) = \frac{1 + s\tau_z}{s(C_z + C_p) \cdot [1 + s\tau_p]}$$
$$\text{with } \tau_z = R_zC_z \text{ and } \tau_p = R_z(C_z^{-1} + C_p^{-1})^{-1} \tag{2.40}$$

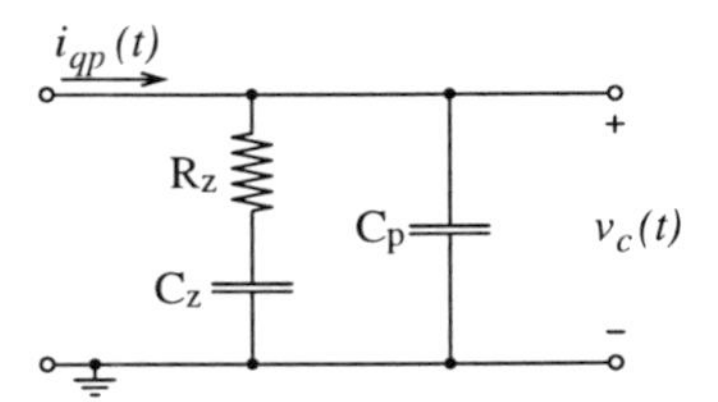

Figure 2.16. Loop filter impedance for a 3-rd order, type-2 charge-pump PLL

The open loop gain equals

$$GH(s) = \frac{I_{qp} \cdot K_{vco}}{2\pi \cdot N} \cdot \frac{1 + s\tau_z}{s^2(C_z + C_p)[1 + s\tau_p]} \tag{2.41}$$

which has a crossover frequency of

$$\omega_c = \frac{I_{qp} \cdot K_{vco} \cdot R_z}{2\pi \cdot N} \cdot \frac{C_z}{C_z + C_p} \tag{2.42}$$

The placement of the extra pole is again a compromise between the intended reference frequency reduction, which requires a low value, and the steady-state and dynamic response of the loop which should not be significantly different from a second-order loop with the same ω_c and τ_z. This last requirement implies the extra pole is placed far to the right with respect to the cross-over frequency. As a general rule of thumb, the zero can be placed at one third of the crossover frequency, and the pole at three times the crossover frequency. This placement guarantees a sufficient phase margin for the loop stability.

The resulting frequency ripple is suppressed with respect to a second-order loop by a factor $\omega_c\tau_p$ [Meyr 90].

2.4.3 Voltage-Controlled Oscillator

Different applications usually require different specifications for the VCO. These requirements are often in conflict with one another, and therefore a compromise is needed. Some of the more important VCO specifications are

- Phase stability : the output spectrum of the VCO should approximate as good as possible the theoretical Dirac-impulse of a single sine wave.
- Electrical tuning range : the VCO must be able to cover the complete required frequency band of the application, including initial frequency offsets due to process variations.

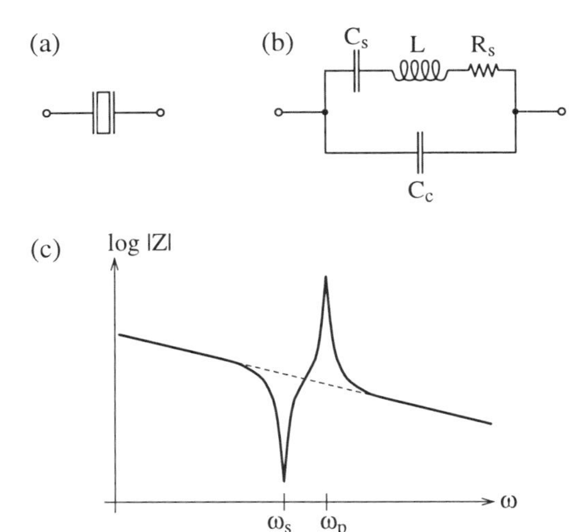

Figure 2.17. An oscillator crystal : (a) Symbol; (b) Electrical model; (c) Impedance

- Tuning linearity : to simplify the design of the PLL, the VCO gain K_{vco} should be constant.
- Frequency pushing : the dependency of the center frequency on the power supply voltage (in $[MHz/V]$).
- Frequency pulling : the dependency of the center frequency on the output load impedance.
- Low cost.

We will now discuss shortly the most often used oscillator types.

2.4.3.1 Crystal Oscillators. The best known and also most stable oscillator is the one based on the resonance characteristic of a crystal (Xtal). The symbol of a crystal is shown in figure 2.17(a), and its electrical model in figure 2.17(b). Typical values for the elements are e.g. : $C_c = 100\ pF$, $C_s = 100\ fF$, $L = 0.1\ H$, $R_s = 10\ \Omega$.

As can be seen from the plot of the impedance of a crystal versus frequency in figure 2.17(c), there are two resonance frequencies. At the series resonance frequency the impedance becomes almost zero. This happens when the negative impedance of the capacitor C_s exactly cancels the positive impedance of the inductor L. So ω_s is given by :

$$\omega_s = \frac{1}{\sqrt{L \cdot C_s}} \tag{2.43}$$

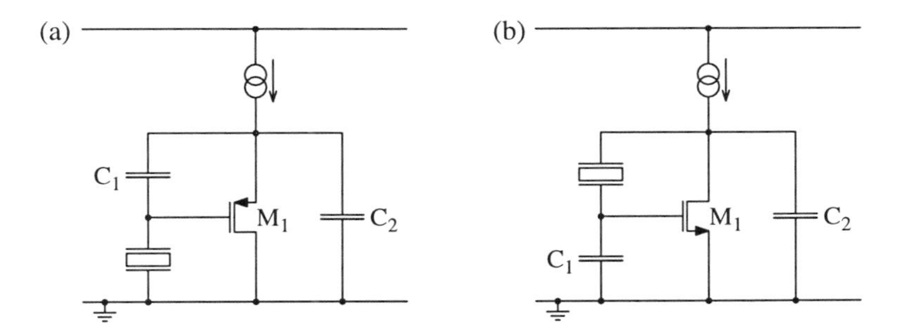

Figure 2.18. Crystal oscillator circuits : (a) One-pin oscillator; (b) Pierce oscillator

Parallel resonance occurs when the impedance in the left branch (with C_c) is exactly the opposite of the impedance in the right branch (with C_s and L). The total impedance is then infinite and ω_p is given by :

$$\omega_p = \frac{1}{\sqrt{L \cdot C_s}} \cdot \sqrt{1 + \frac{C_s}{C_c}} \tag{2.44}$$

Due to the large ratio between C_s and C_c, ω_s and ω_p are very close together. This large ratio is also the reason why crystal oscillators achieve such very low phase noise, as will be explained in chapter 3.

A lot of effort has been spent to make integrated Xtal oscillators, i.e. having a crystal as the only external component [Vitto JSSC88, Santo JSSC84, Huang JSSC88, Nordh CAS90, Laund UWRF92]. Most designs are based on the general theory developped in [Vitto JSSC88] that allows accurate linear and nonlinear analysis of the crystal oscillator circuit. Using a single transistor as the active element of the oscillator, only two possible circuit schematics are possible, both of which are shown in figure 2.18. The circuit in figure 2.18(a) requires only one additional pin for the crystal, as the other side can be connected to ground. This is the preferred circuit for e.g. the clock oscillator in large digital ICs, where the total number of pins must be kept as small as possible. The Pierce oscillator shown in figure 2.18(b) needs two separate pins to connect the crystal, but provides a better frequency stability [Vitto JSSC88].

Crystals come in many varieties, each with different specifications of center frequency, frequency variation, stability, etc. The most common frequencies are of the order of 10 MHz, but crystals using a harmonic overtone of the fundamental resonance frequency achieve several hundred MHz. Their low noise and good stability makes them the first choice for generating the reference signal in a frequency synthesizer.

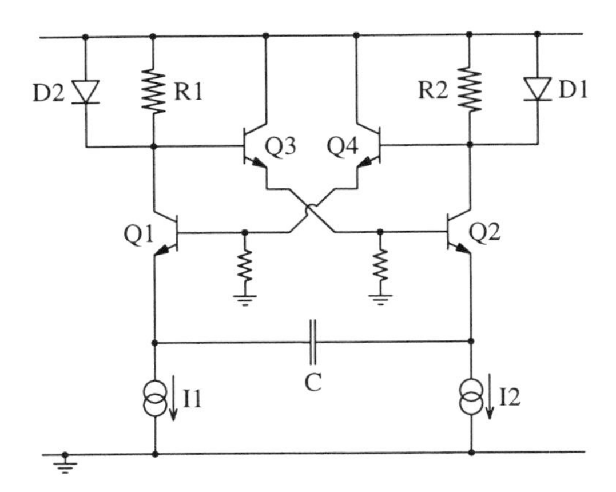

Figure 2.19. Emitter-coupled multivibrator

The crystal properties drift over time and temperature. To control the temperature drift, Temperature-Controlled Crystal Oscillators (TCXO) put a miniature heater around the crystal to keep its temperature constant. Voltage-Controlled Crystal Oscillators (VCXO) put a tunable capacitor in parallel with the crystal to vary the value of C_c. This way a small tuning range (usually in the order of 0.1%) can be implemented.

2.4.3.2 Relaxation Oscillators. Another category of oscillators that is often used for monolithic integration, and that is capable of high speeds, are relaxation oscillators. The basic circuit is the emitter-coupled multivibrator [Gray 93, Abidi JSSC83]. A general circuit diagram is shown in figure 2.19. Transistors $Q1$ and $Q2$ are alternately switched on and off, and the timing capacitor C is charged and discharged with current sources $I1$ and $I2$. Transistors $Q3$ and $Q4$ are level shifting emitter followers, and diodes $D1$ and $D2$ define the voltage swings at the collectors of $Q1$ and $Q2$. A triangle wave is obtained across the capacitor and square waves at the collectors of $Q1$ and $Q2$.

By changing the charging current, very wide tuning ranges are possible. Several changes to this scheme can be made, e.g. implementation in CMOS [Banu JSSC88]. The phase noise generation theory can be found in [Abidi JSSC83, Sneep JSSC90]. The resulting equation takes the form [Sneep JSSC90]

$$\mathcal{L}\{\Delta\omega\} = a \cdot \frac{kT\,R_n}{V_A^2} \cdot \left(\frac{\omega_0}{\Delta\omega}\right)^2 \tag{2.45}$$

with a a factor depending on the noise generation mechanism studied, R_n an equivalent noisy resistor, V_A the voltage amplitude of the signal, ω_0 the oscillation frequency and $\Delta\omega$ the frequency offset. The only way to lower the phase noise is to lower

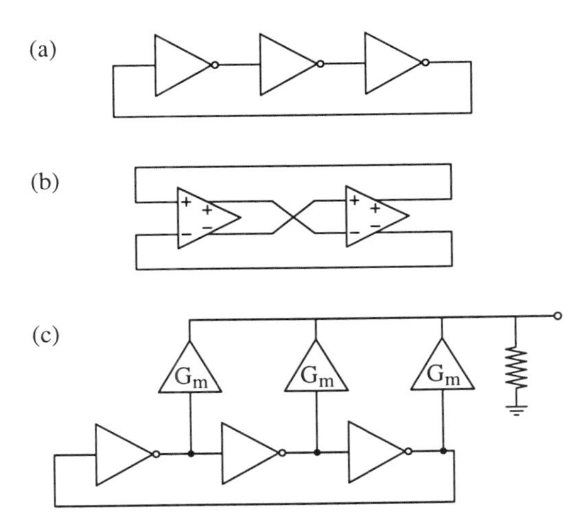

Figure 2.20. Ring oscillator : (a) three-stage; (b) differential two-stage; (c) 1-stage delay

the resistance value, but this inherently implies a larger power consumption. Typical specs for relaxation oscillators are a maximum oscillation frequency in the order of a few 100 MHz, and a phase noise value of -120 dBc/Hz at 1 MHz offset [Sneep JSSC90, Kato JSSC88, Liu JSSC90].

2.4.3.3 Ring Oscillators. A very important integrated oscillator configuration is the ring oscillator. Due to its simplicity and its ease in IC-integration, it is used in a lot of PLL frequency synthesizer and clock recovery designs [Chen ISSCC93, Banu ISSCC93, Razav ISSCC94b, Pottb JSSC94]. The periodic signal is generated by a ring of inverters. The simplest configuration, using only three inverters, is shown in figure 2.20(a). The oscillation period will be $2n \cdot T_d$, with n the number of inverters in the ring and T_d the delay of one inverter.

Tuning is possible by varying the current the inverters use to charge their load capacitance. In that case, we can talk of a Current-Controlled Oscillator (CCO). Another interesting tuning strategy is proposed in [Enam JSSC90]. A voltage divider allows to change continuously from a ring containing 5 inverters to a ring containing 9 inverters. That way the inverters used can have the basic 2-transistor inverter topology, since the charging current doesn't need to be regulated. Higher oscillation frequencies are possible, since inverter delay T_d will be smaller. The effect of this on the phase noise performance has not been investigated.

Normally the number of inverters in the loop must be odd and larger than one. So the minimum number is three, which poses a limit on the maximum frequency. Differential inverters allow the use of only two inverters in the ring, as shown in figure 2.20(b). But since the differential topology will be more complex than the normal one, it is not sure that a speed enhancement is obtained. However, by combining the three available signals of a three-stage oscillator as shown in figure 2.20(c), an output signal can be obtained that has an oscillation period of a single-stage ring [Razav ISSCC94b]. A distinct advantage of even-stage differential ring oscillators in the intrinsic availability of quadrature signals [Abidi CICC94].

Recent submicron realizations of ring oscillators have demonstrated their capability of achieving GHz oscillation frequencies [Abidi CICC94, Thams CICC95]. The only aspect that impedes their use in mobile communications ICs is their rather high phase noise. The switching action introduces a lot of disturbances in the oscillator, so ring oscillators usually have bad phase noise characteristics. They also cannot benefit from the bandpass characteristic of e.g. an LC-tank to reduce phase noise. A general discussion of ring oscillator phase noise is given in [Weiga ISCAS94] and [Razav JSSC96]. The resulting equation has the same form as (2.45). So for these types of oscillators we can also conclude that the phase noise can be lowered by increasing the power consumption linearly. Typical noise values are e.g. -83 dBc/Hz at 100 kHz offset from a 900-MHz carrier [Thams CICC95], or -94 dBc/Hz at 1 MHz offset from a 2.2-GHz carrier [Razav JSSC96].

2.4.3.4 LC-Oscillators. Both the relaxation oscillator and ring oscillator are capable of achieving an operating frequency in the GHz range when implemented in an advanced submicron CMOS technology, and they require no external components. But their phase noise behavior is generally not good enough to be applicable in a mobile cellular telephone system. Lowering the phase noise to an acceptable level requires too much power. The only other monolithic oscillator type is the one based on the resonance of an inductor and a capacitor connected in a loop, the LC-oscillator. Since the oscillation frequency is determined by passive elements, it is expected that these oscillators will have a very pure spectrum. A detailed analysis of their phase noise behavior is given in chapter 3. Depending on the exact implementation, and on the quality of the passive elements, typically a 20-dB better phase noise is obtained over ring and relaxation oscillators. Also high speed operation is possible due to the simple working principle.

The biggest difficulty in realizing a monolithic LC-tuned VCO is of course the realization of the inductor. Most designs use external elements, either a real inductor [Soyue JSSC89], or an off-chip transmission line acting as an inductor [Wang EL92, Wang ISSCC94]. Active inductors are made out of a capacitor and some active elements [Zhang UWGW93, Wang JSSC90], which are all available in IC-technology. They have the disadvantage of consuming a lot of power, and generating noise, as will

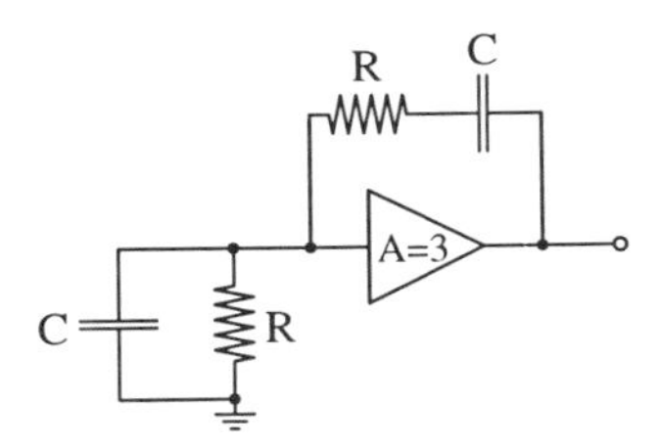

Figure 2.21. Wien-bridge oscillator

be shown in chapter 3. On-chip passive inductors can be realized using bonding wire inductors, which are presented in chapter 4, or planar inductors, which are discussed in chapter 5.

2.4.3.5 OTA-C Oscillators. Positive feedback that creates the instability necessary for oscillation can also be made using only capacitors and transconductors, and sometimes also resistors. The most classical example is the Wien-bridge oscillator, which is shown in figure 2.21, that can be implemented using Operational Transconductance Amplifiers (OTAs), capacitors and resistors [Senan EL89a, Senan EL89b]. A general theory for OTA-C oscillators was developed in [Rodri CAS90]. The class of OTA-C oscillators without resistors can be analyzed as being LC-tuned oscillators with active inductors. For example, in [Senan EL89b] the elements forming the amplifier, the capacitor and the active inductor can clearly be distinguished.

2.4.3.6 Other Configurations. Apart from the oscillator types listed in the previous sections, several other high-frequency oscillator types exist. They are not explained here, since they are not suitable for IC-implementation. Amongst these types are Dielectric Resonance Oscillators (DRO), Yttrium Iron Garnet (YIG) oscillators, etc.

2.4.4 Frequency Divider

The most apparent feature that differentiates the PLL frequency synthesizer from other phase-locked loops is the frequency divider. The several implementation options and their consequences are discussed in this section.

A distinction can be made between two main types of circuits : synchronous and asynchronous [Egan 81]. In a synchronous circuit, every flipflop is triggered by the divider input signal (clock) itself. In an asynchronous divider, the input signal triggers only the first flipflop, this one triggers the second, etc. Because of the direct input from the clock signal, without any delays through other stages, synchronous dividers achieve a complete transition faster.

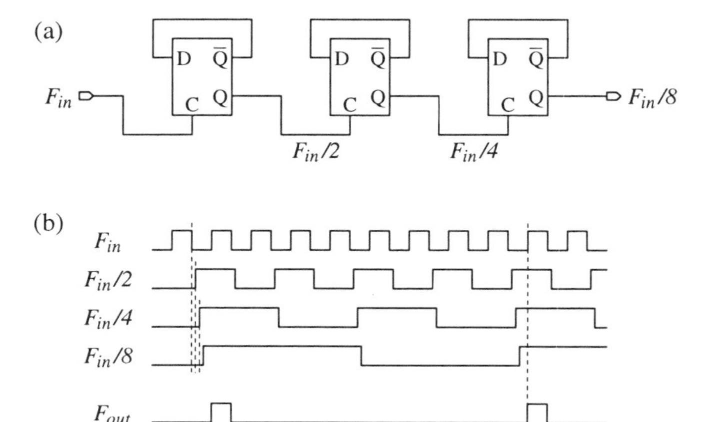

Figure 2.22. (a) Asynchronous frequency divider; (b) Waveforms

Figure 2.22 shows an asynchronous three-stage divider. Every stage performs a divide-by-2 function, which is the input of the next stage. Note how the output of each divider lags the output of the previous one. This makes the output of the third stage to be asynchronous with respect to the input clock. If you combine the four signals in an AND-gate to obtain a signal F_{out}, this will be, apart from the delay in the AND, synchronized to the clock.

Synchronous dividers can be made using JK-flip-flops, as shown in figure 2.23. All stages change almost simultaneously on the edge of the input clock. This will make the last stage respond more rapidly, but care must be taken when combining several of the intermediate frequencies in an output AND-gate. A race problem can occur creating unwanted spikes when some signals change faster or slower than the others.

2.4.4.1 Programmable Dividers or Counters. A programmable counter is used to obtain a variable, controllable divide ratio. A certain preset number can be loaded in the counter, which starts counting input pulses until the end number is reached and an overflow signal is generated. It is then reset to the original preset number and starts counting again. The division ratio then equals $2^n - P$, with n the number of bits of the counter and P the preset number.

Another implementation starts counting input clock cycles from zero, until a certain value P is reached and a reset signal is generated. The counter result is then reset to zero and the counting process restarts. The divide ratio then equals P.

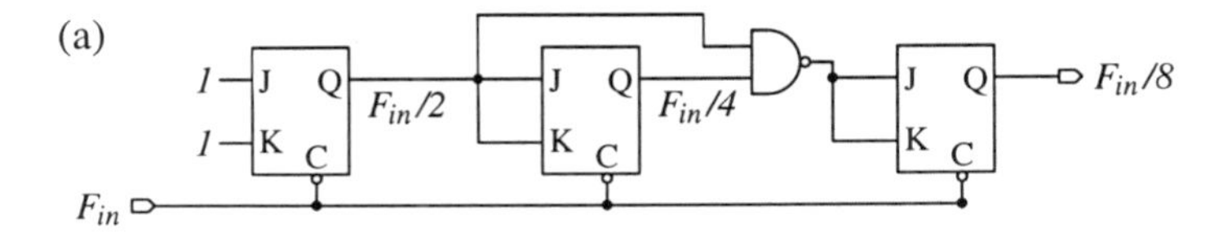

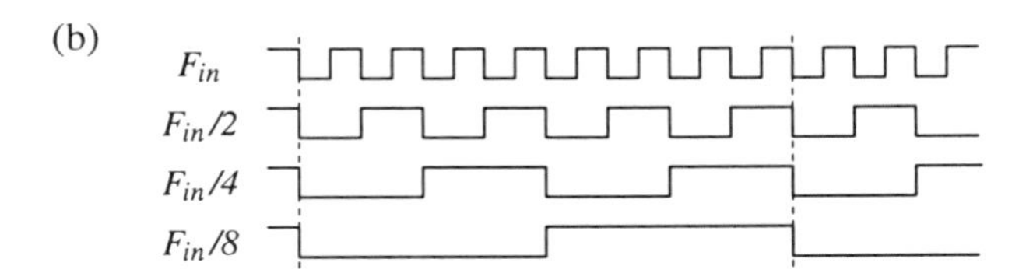

Figure 2.23. (a) Synchronous frequency divider; (b) Waveforms

2.4.4.2 Prescalers. When the divider input frequency is too high to permit proper operation of a programmable divider or counter, a *prescaler* can be used. A prescaler divides by a fixed ratio, and can therefore operate at higher frequencies because it does not have to allow for the delays involved in counting and (p)resetting. A few high-speed prescaler stages permit lowering the required speed for the subsequent counter stages. A disadvantage is that, for a given frequency resolution (channel spacing), the reference frequency must be lowered. If the prescaler division ratio is N_p, the smallest change in synthesized frequency is $\Delta f = N_p \cdot F_{ref}$. So the reference frequency must be a factor N_p lower than the channel spacing. This implies a lower loop bandwidth, which is often undesirable.

A solution to this frequency resolution problem, is the use of a Dual-Modulus Prescaler (DMP). This circuit extends a fixed-ratio prescaler with some extra logic that enables it to divide by N_p or by $N_p + 1$. The speed decrease due to this extra functionality is not necessarily negligible, but can be kept limited. An example of a synchronous divide-by-4/5 circuit is shown in figure 2.24. The critical path consists of a NAND gate and a D-flipflop delay.

The use of such a DMP in a full divider that can handle all integer ratios is shown in figure 2.25. It incorporates a DMP, a Programmable counter (P) and a so-called Swallow counter (S). The DMP divides the (high) input frequency by $N_p + 1$. The S-counter counts the DMP output pulses, until a number S is reached. It then changes the DMP modulus control, which starts dividing by N_p. The DMP output pulses are also counted in the P-counter. If this last counter has counted P pulses, it resets itself and the S-counter. Both counters restart counting from zero, while the DMP divides again by $N_p + 1$. So during one output period, the DMP has divided S times by $N_p + 1$

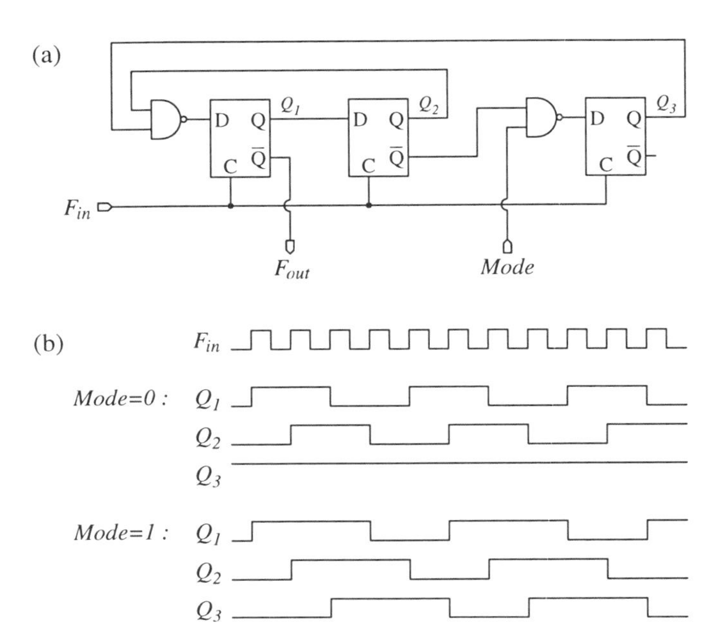

Figure 2.24. (a) Dual-modulus divide-by-4/5 prescaler; (b) Waveforms

and $P - S$ times by N_p, which yields an overall division ratio of

$$N = P \cdot \left[(N_p + 1) \cdot \frac{S}{P} + N_p \cdot \frac{P - S}{P} \right] = P \cdot N_p + S \tag{2.46}$$

If one wants to be able to obtain the complete range of integer numbers, S must be variable from 0 to $N_p - 1$. And for proper reset by the P-counter, P must always be larger than the largest value of S, or $P \geq N_p$. So the smallest obtainable division ratio is $N_{min} = N_p^2$. For a given minimum synthesized frequency, and a given frequency resolution, this puts a limit on the maximum prescaler division ratio possible.

2.4.4.3 Fractional-N Synthesis. The low reference frequency is a general problem in a lot of PLLs. It is required by several factors. First, there is the channel spacing which is e.g. 200 kHz in GSM. Since the frequency divider modulus N can only be programmed to integer values, this implies that the reference frequency used in the synthesizer must be as low as or lower than 200 kHz. The use of a prescaler decreases the reference frequency again by a factor N_p. And a dual-modulus prescaler has a certain minimum division number, so it is not guaranteed that it can use a higher reference frequency than a simple prescaler.

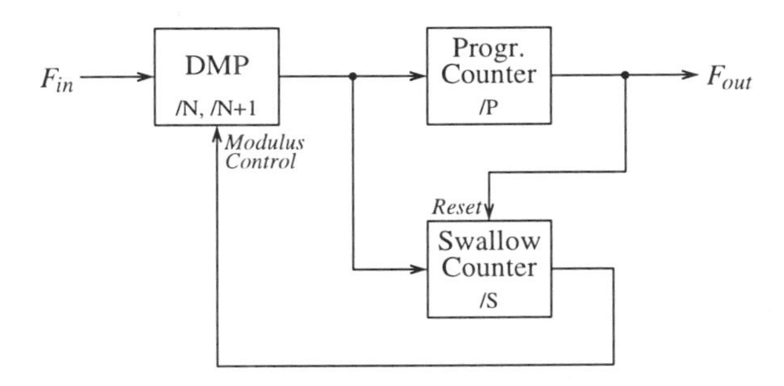

Figure 2.25. Frequency divider with dual-modulus prescaler and pulse- and swallow-counters

Two problems are created by a too low value of F_{ref}. First, to assure a proper linear operation of the phase detector, the loop bandwidth must be limited to approximately one tenth of that value. This is an important restriction. Normally, a larger loop bandwidth would be desirable, because this allows faster switching times and more suppression of the VCO noise. Secondly, because of the high value of N, the noise of the reference signal, the phase detector and the loop filter will become rather high, since these noise sources are all multiplied by a factor N before they show up in the output spectrum (see section 2.3.1). If the LO carrier is 2 GHz, and the reference frequency is 200 kHz, N equals 10^4. So the reference and PD noise are deteriorated by 80 dB.

Fractional-N frequency synthesizers solve this problem partially by switching very fast between two divider moduli N and $N+1$. This is shown in figure 2.26. The accumulator increases its value with an input number n every clock cycle of the reference frequency F_{ref}. Its overflow indicator is used as the modulus control input to the divider. If the accumulator overflows, the division factor is $N+1$, otherwise it equals N. So on the average, the divider divides n times by $N+1$ and $1-n$ by N, resulting in a synthesized frequency of

$$F_{out} = [n \cdot (N+1) + (1-n) \cdot N] \cdot F_{ref} = [N+n] \cdot F_{ref}. \qquad (2.47)$$

This means that also non-integer division factors can be realized in the prescaler, and the above-mentioned limitation on the reference frequency is not applicable here. Of course, this technique also has disadvantages. The most important one is the generation of spurs in the output spectrum due to pattern noise in the overflow signal. This can be better understood if the accumulator is regarded as a first-order Sigma-Delta ($\Sigma\Delta$) modulator [Riley JSSC93]. As the input to this modulator is a DC signal, the quantization noise is not randomized, and the output contains many spurious signals. As the pattern noise of higher-order $\Sigma\Delta$-modulators is much lower, the accumulator

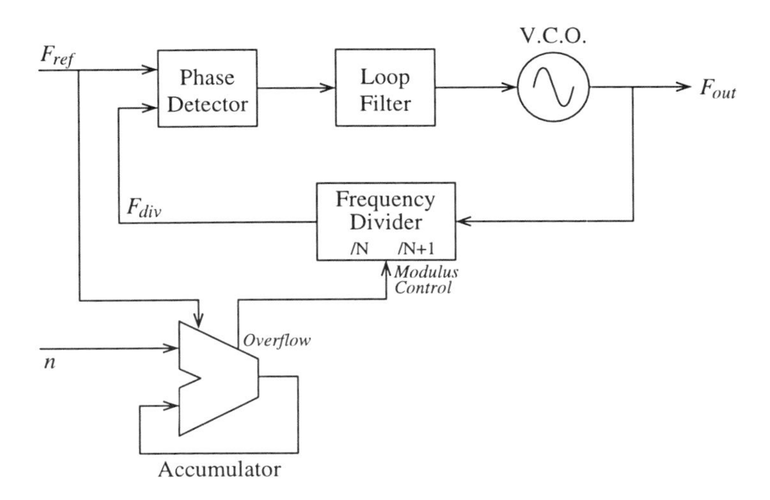

Figure 2.26. PLL frequency synthesizer with fractional-N division

can be replaced by a higher-order digital Sigma-Delta ($\Sigma\Delta$) modulator to limit the frequency spurs [Riley JSSC93, Mille TIAM91, Marqu JWN97].

2.5 CONCLUSION

In this chapter we have given an introductory study of PLL frequency synthesizers. Two classes of performances specify the PLL. The output noise is very important in LO synthesizers, and is determined by the noise sources present and by their transfer functions to the output. VCO noise is high-passed to the output, so for frequencies larger than the loop bandwidth the output noise is completely determined by the quality of the VCO. Reference noise is low-passed and multiplied by the division modulus N. The transient performances of the PLL, such as tracking and acquisition, are determined by the order and type of the loop, and most important by the loop bandwidth.

The different loop building blocks have been discussed. Some possible implementation options for each component are given, together with the expected advantages and disadvantages. To summarize, we can make a prediction on the circuits to be used in a fully integrated CMOS LO synthesizer for mobile communication systems.

- The building block that will require the most effort in a fully integrated design is probably the VCO. Of the several types presented, only the LC-tuned oscillator seems a feasible option. Most oscillators cannot handle the high frequencies required, and those who can, i.e. the relaxation and the ring oscillator, suffer from too much phase noise. The LC oscillator offers the distinct advantage that phase noise as well as power consumption improve with the inductor's quality factor, as will be demonstrated in chapter 3 which investigates the general theory of phase noise in LC oscillators. So the only problem left is to design an integrated inductor with low losses. Because this is of course easier said than done, a large part of this work in contributed to inductor design and LC VCO design. Chapters 4 and 5 discuss two classes of implementation, i.e. bonding wire inductors and planar integrated spiral inductors.

- The other block that operates at the full frequency, and hence requires special attention in a CMOS design, is the frequency divider. Complicated circuits such as programmable dividers and counters are not capable of GHz operation. So the inclusion of a prescaler will be necessary. Chapter 6 investigates this.

- As for the low-frequency PLL blocks, it seems the most straightforward implementation is the one with a phase-frequency detector and a charge pump loop filter. The PFD circuit of figure 2.11 is easily realized in CMOS, and the charge pump allows a possible passive realization of the loop filter.

- Of course, the determination of the order and type of the PLL, of the loop bandwidth, of the reference frequency and the associated necessity of fractional-N synthesis, and of the several gain factors such as VCO gain K_{vco}, division modulus N, charge pump current I_{qp}, loop filter components R_z, C_z and C_p, etc. still requires a lot of investigation. They will partly be determined by the circuit realizations of the blocks, by the transient requirements of the synthesizer, and by the noise specifications of the several blocks. Chapter 7 deals with these aspects in detail.

3 VOLTAGE-CONTROLLED OSCILLATOR PHASE NOISE

3.1 INTRODUCTION

In this chapter the theoretical calculation of the phase noise of an LC-tuned oscillator is developed. As we have already stated, this is the only type of integrated oscillator that will be capable of achieving the required high frequency and low noise at a reasonable power consumption.

The first section discusses the general LC-oscillator theory. The most important part of the oscillator is the inductor. The reason for this is the fact that inductors are not standard available components of a CMOS IC technology. Several parasitic effects appear, severely influencing the performances of the inductor and hence of the oscillator. The different options are presented shortly.

The several noise contributions of the amplifier, the resistors and the inductor to the output spectrum are explained in sections 3.3, 3.4 and 3.5. The concepts of *effective resistance* and *effective capacitance* are introduced, which allow an instant evaluation of every LC-tank concerning its noise and required power. Also improvements to the standard LC-tank are presented, which allow a trade-off between noise and power. This is shown in section 3.6.

Section 3.7 handles some other noise sources, which contribute to the output phase noise due to FM upconversion by non-linear effects in the oscillator's amplifier and tuning capacitor.

Finally, section 3.8 gives an overview of the current status of state-of-the-art VCOs published in open literature recently, and how the VCO designs reported in this work relate to them.

3.2 OSCILLATOR THEORY

An LC-tuned oscillator is basically a feedback network as shown in figure 3.1. Stable oscillation will occur at the frequency at which the loop transfer function $GH(s)$ is exactly equal to one. This is known as the Barkhausen criterion [Grebe 84]. The oscillation frequency can easily be found, because the imaginary part of $GH(s)$ has to be zero. We can give an intuitive explanation for this. Suppose we excite the circuit such that there is a certain signal at a certain node in the circuit. If the imaginary part of the loop transfer function is zero, than this signal will go round the loop and come back *in phase* where it is added to the original signal. If the loop transfer function is smaller than one, the new signal will be smaller than the original and the excitation will die. If the loop transfer function is larger than one, the new signal will be larger than the original and the excitation will be amplified. This is the condition for oscillation.

In figure 3.2 the basic LC-tuned oscillator that will be used in the calculations in the following sections is shown. It is based on the infinite impedance of the parallel connection of a capacitor and an inductor at their resonance frequency. Clearly the circuit is unstable, since the feedback over the transconductor G_M is positive. All parasitic resistances are also shown. A series resistance R_c is associated with the capacitor C, and a series resistance R_l with the inductor L. The output resistance of the transconductor, and the parallel resistances across C and L, are represented by R_p.

An important restriction made here is the assumption of a linear circuit. This means the amplitude of the oscillation is limited such that the amplifier used remains linear. This has consequences on the noise calculations in the following sections. In a non-linear circuit, baseband noise is upconverted to the oscillation frequency due to non-

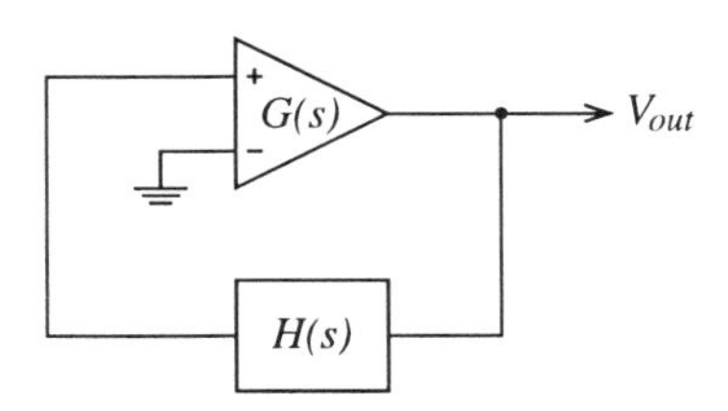

Figure 3.1. General feedback network

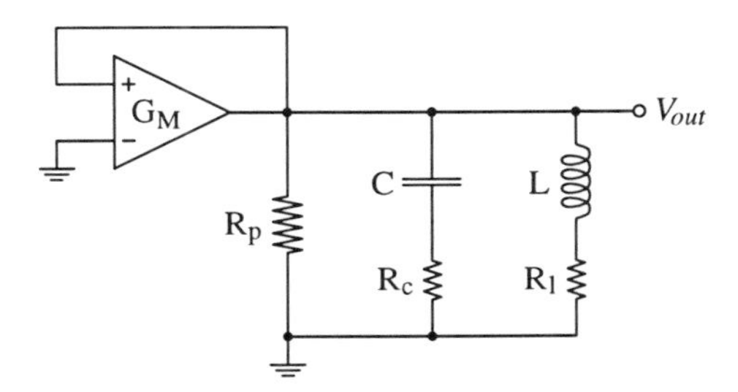

Figure 3.2. Basic LC-tuned oscillator configuration

linear effects. So low-frequency $1/f$ noise plays an important role in the close-to-the-carrier output spectrum. This will be dealt with in section 3.7. In a linear circuit, only high-frequency thermal noise will be important.

Since we first only consider linear circuits, we will assume that the forward gain G (i.e. the transconductance G_M) will be such that the open loop gain $GH(s)$ at resonance is exactly equal to one. In a real, non-linear circuit, this will not be the case. A safety factor of two to three is usually included to ensure reliable startup. If the oscillator is equipped with an Automatic Gain Control (AGC) circuit, the amplification will be lowered as soon as the desired amplitude of oscillation is achieved. The amplitude can also be limited by non-linear effects, e.g. clipping at the power supplies or a limited input range of the amplifier used. In that case, the non-linearities will ensure that there is some sort of average or effective amplification such that $GH(s)$ becomes equal to one.

As already mentioned several times, the key element in the realization of a good on-chip LC-tank is the design of the inductor. Capacitors are readily available in all IC technologies. If capacitors with low series resistance are needed, metal/metal capacitors can be used. A typical value of the sheet resistance of a metal layer in an IC technology is 50 $m\Omega$ per square, which allows the realization of capacitors with a very high quality factor [Soyue JSSC96a]. In most cases, the performance of the LC-tank will be determined by the performance of the inductor. A standard IC technology that allows the realization of an inductor doesn't exist, so some tricks have to be used. This severely decreases the performance of the integrated inductor, e.g. limited operating frequency due to parasitic capacitance or low quality factors. The remaining part of this section describes some design options that can be taken. First, the definition of an often used term, i.e. the quality factor of an element, is given. Then active integrated inductors are dealt with : active elements are used to transform the impedance of a capacitor into the impedance of an inductor. Next the passive implementations of integrated inductors are discussed shortly. They can be bonding wire inductors or the common known spiral or planar on-chip inductors. A thorough analysis of both these types of inductors and their application in VCO design will be given in chapters 4 and 5, respectively.

3.2.1 Q : the Quality factor

Some confusion might arise when discussing quality factors. Therefore, a general definition is given here :

$$Q = 2\pi \cdot \frac{\text{Peak energy stored}}{\text{Energy loss per cycle}} \tag{3.1}$$

Using this formula, e.g. the well-known quality factor of an LC-tank with a parallel resistor R_p can be calculated. Suppose a sinusoidal voltage $V(t) = V_A \cdot \sin(\omega t)$ is applied to the circuit. At times t when the applied voltage reaches its maximum, the voltage across C equals V_0 and the current through L equals zero. At that time the energy stored in the capcitor is at a maximum, and the energy in the inductor is zero. So the numerator of (3.1) is :

$$E_{peak} = \frac{C \cdot V_A^2}{2} \tag{3.2}$$

Energy is lost in the circuit through the resistor R_p. This amount of energy lost can be calculated to be :

$$E_{loss} = \int_0^{2\pi/\omega} \frac{[V_A \cdot \sin(\omega t)]^2}{R_p} \cdot dt = \frac{\pi \cdot V_A^2}{\omega \cdot R_p} \tag{3.3}$$

Combining these two equations we can easily find the quality factor of an LC-tank with parallel resistance.

$$Q_{L,C,R_p} = 2\pi \cdot \frac{\frac{C \cdot V_A^2}{2}}{\frac{\pi \cdot V_A^2}{\omega \cdot R_p}} = R_p \cdot \omega C \tag{3.4}$$

This quality factor is frequency dependent. At the LC-tank's resonance frequency $\omega_0 = 1/\sqrt{LC}$ this results in the well-known expression

$$Q_{L,C,R_p}(\omega_0) = R_p \cdot \sqrt{\frac{C}{L}} \tag{3.5}$$

Quality factors for single inductors or capacitors can also be calculated using (3.1). Table 3.1 gives the results.

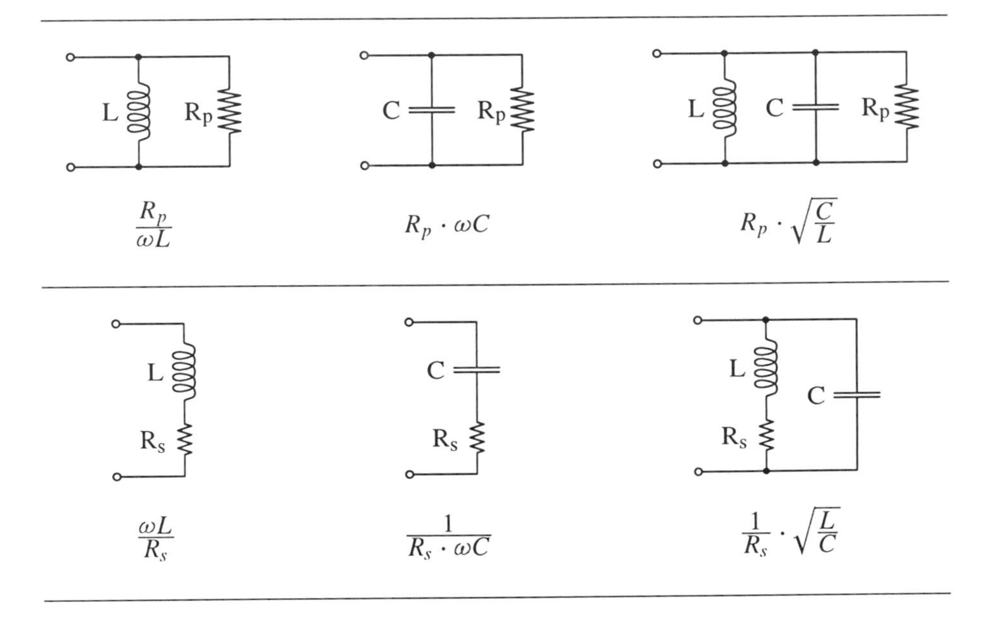

Table 3.1. Quality factors for several circuits

3.2.2 Active Inductors

3.2.2.1 Circuit implementation. An often used way to integrate an inductor is to simulate one with active elements. These active elements can modify the impedance of a capacitor such that the input impedance of the circuit resembles that of an inductor. In general, a gyrator and a capacitor are required. Figure 3.3(a) shows the symbol of a gyrator. The most often used implementation, i.e. the anti-parallel connection of two transconductors [Wang JSSC90], is shown in figure 3.3(b).

As shown in figure 3.3(c), the impedance of a capacitor seen through a gyrator looks inductive. It can easily be calculated that the inductance value is given by :

$$L_{act} = \frac{C_L}{g_{mi} \cdot g_{mv}} \tag{3.6}$$

The frequency capability of these active implementations of inductors can be quite high. A GaAs implementation of a floating active inductor [Zhang UWGW93] can operate up to half the cut-off frequency of the transistors. Therefore, a submicron CMOS version of this circuit is expected to be able to operate at 1 or 2 GHz.

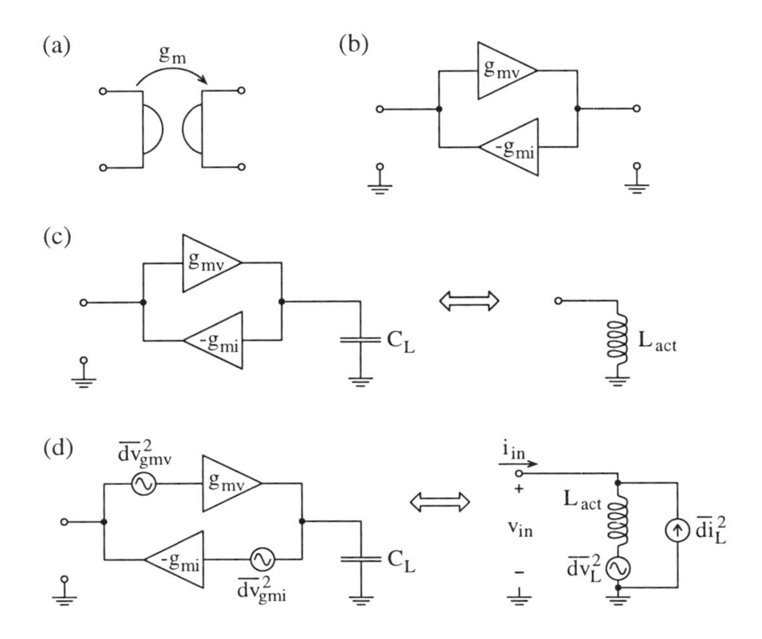

Figure 3.3. Active inductor : (a) Gyrator symbol; (b) Implementation; (c) Simulation of an inductor; (d) Equivalent noise sources

3.2.2.2 Noise in Active Inductors. The biggest problem associated with active inductors is their noise contribution to the circuit. Noise is generated by both transconductors in the gyrator. The transconductors g_{mi} and g_{mv} have been given a subscripts i and v. This subscript refers to whether they generate a current noise source (i) or a voltage noise source (v). Figure 3.3(d) shows the equivalent noise sources in an active inductor.

Comparison of the noise in both circuits of the figure can be done in the following way. Assume that the input node of the inductor is shorted to the ground, so v_{in} equals zero. The input noise current of the right hand side of figure 3.3(d) then equals

$$\overline{di}^2_{in(right)} = \overline{di}^2_L + \frac{1}{(sL_{act})^2} \cdot \overline{dv}^2_L \tag{3.7}$$

The same calculation can be made for the left hand side of figure 3.3(d), which results in

$$\overline{di}^2_{in(left)} = g^2_{mi} \cdot \left[\left(\frac{g_{mv}}{sC_L} \right)^2 \cdot \overline{dv}^2_{gmv} + \overline{dv}^2_{gmi} \right] \tag{3.8}$$

Since both equations are equivalent, and using (3.6), the equivalent noise sources of an active inductor can be expressed as

$$\begin{aligned} \overline{di_L^2} &= g_{mi}^2 \cdot \overline{dv_{gmi}^2} \\ \overline{dv_L^2} &= \overline{dv_{gmv}^2} \end{aligned} \tag{3.9}$$

These expressions can be further clarified by substituting an expression for the transconductor equivalent input noise source

$$\overline{dv_{gm}^2} = 4kT \cdot \frac{F}{g_m} \cdot df \tag{3.10}$$

The factor F represents the excess noise factor of the transconductor, and is dependent on its particular implementation. For a single MOS transistor, a factor of $2/3$ is usually used. Equation (3.9) now reduces to

$$\begin{aligned} \overline{di_L^2} &= 4kT \cdot F_i \cdot g_{mi} \cdot df \\ \overline{dv_L^2} &= 4kT \cdot \frac{F_v}{g_{mv}} \cdot df \end{aligned} \tag{3.11}$$

with F_{Li} the excess noise factor of G_{mi} and F_{Lv} of G_{mv}.

These formulas explain the subscripts used for the transconductances g_{mi} and g_{mv}, since their associated noise sources are responsible for either an equivalent current noise source, or an equivalent voltage noise source in the active inductor. In section 3.4 it will be shown that these noise sources have a very bad effect on the phase noise of an LC-tuned oscillator. Therefore, passive implementations of integrated inductors are necessary.

3.2.3 Passive Inductors

Passive on-chip inductors are mostly realized using a spiral-shaped metal connection, routed in one or more of the standard available routing levels. An extensive calculation method for planar rectangular micro-electronic inductors is given in [Green TPHP74]. These inductors suffer from several losses, such as the resistive losses in the metal connections and in the underlying conductive silicon substrate and the parasitic capacitance to the substrate. A thorough analysis of planar inductors is presented in chapter 5.

An alternative solution that suffers less from the above-mentioned problems takes advantage of the parasitic inductance that is usually associated with bonding wires in IC packages. A rule of thumb for this inductance is 1 nH per mm length. So inductance values of several nH can be realized. In combination with a 4 pF capacitor, only 6 nH is needed to create an LC-tank oscillating at 1 GHz. The series resistance

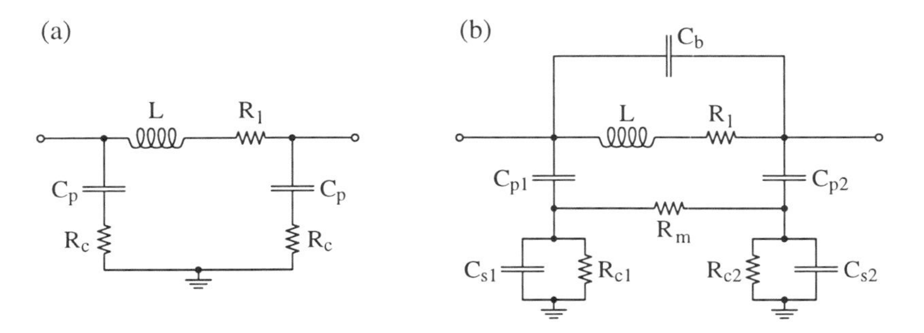

Figure 3.4. Equivalent circuit of a passive integrated inductor : (a) simple; (b) complete

of a bonding wire is very low, so high quality factors can be accomplished. Also parasitic capacitance to the substrate is limited to the capacitance of two bonding pads, so the self-resonance frequency will be high. These aspects illustrate the feasibility of bonding wires in LC-tuned oscillators, and a thorough analysis is presented in chapter 4.

Generally, passive inductors can be modeled by the simple equivalent circuit shown in figure 3.4(a) [Nguye JSSC90]. The inductor L has a parasitic series resistance R_l and a parasitic capacitance to ground C_p. Sometimes this parasitic capacitance has an associated series resistance R_c. Extensions are to put a bridging capacitor directly across the inductor and its series resistance to account for the fringing capacitances, or to place a parallel resistor to model the magnetic losses. One can also split up the capacitance to ground into a lossless and a lossy part. The resulting circuit is shown in figure 3.4(b). This model can be fitted onto the measured data of almost every integrated spiral inductor [Nikne CICC97].

The only noise sources in a passive inductor are the parasitic resistances, the most important one of which will be R_l. So if the noise of a passive inductor is written in the same form as (3.11), the result is :

$$\begin{aligned} \overline{di_n^2} &= 0 \\ \overline{dv_n^2} &= 4kT \cdot R_l \cdot df. \end{aligned} \tag{3.12}$$

In the next section we will discuss the effect of the passive inductor's parasitic resistance on the oscillator's phase noise.

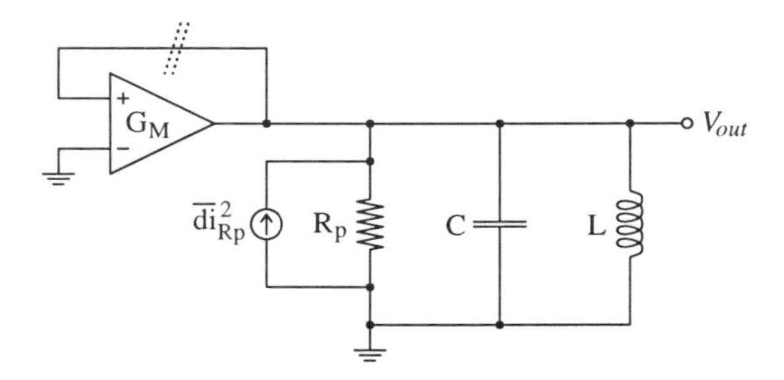

Figure 3.5. Basic oscillator with noisy R_p

3.3 PHASE NOISE OF A BASIC OSCILLATOR WITH PASSIVE INDUCTOR

3.3.1 *Parallel Resistance* R_p

In figure 3.5, the basic oscillator circuit is redrawn, with R_p as the only parasitic effect. A current noise source $\overline{di}^2_{Rp}$ is also included.

As a first step, the oscillation frequency has to be calculated. Therefore the loop transfer function is derived by cutting the loop as shown by the two dotted lines in figure 3.5. The result in this case is

$$T_{loop,R_p}(s) = G_M \cdot \frac{sL}{1 + s\frac{L}{R_p} + s^2 LC} \tag{3.13}$$

The imaginary part of the loop transfer function is equal to

$$\Im\left\{T_{loop,R_p}(\omega)\right\} = G_M \cdot \frac{\omega L \cdot \left(1 - \omega^2 LC\right)}{\left(1 - \omega^2 LC\right) + \omega^2 \cdot \left(\frac{L}{R_p}\right)^2} \tag{3.14}$$

and is zero for

$$\omega_0 = \frac{1}{\sqrt{L \cdot C}} \tag{3.15}$$

This frequency ω_0 is the frequency at which the circuit will oscillate. Now the transconductance G_M, necessary to have a loop transfer function exactly equal to one, can be calculated. We use the symbol G_{M,R_p} because the transconductance value is the one needed to compensate for the losses in the resistor R_p. It is given by

$$G_{M,R_p} \doteq \frac{G_M}{T_{loop}(\omega_0)} = \frac{1}{R_p} \tag{3.16}$$

Then the transfer function from the current noise source $\overline{di}^2_{Rp}$ to the output voltage V_{out} in closed loop operation must be calculated. This results in

$$T^2_{noise,R_p}(s) \doteq \frac{\overline{dV}^2_{out}}{\overline{di}^2_{Rp}}(s) = \left[\frac{sL}{1 - sL \cdot (G_M - G_p) + s^2LC}\right]^2 \tag{3.17}$$

The notation G_p represents the inverse of the resistance R_p. To calculate the phase noise of the oscillator, we have to evaluate this function at a frequency $\omega_0 + \Delta\omega$ that is slightly offset to the center frequency. At the center frequency ω_0 itself, the noise transfer function is infinite, which is the basic reason for oscillation. To render the calculation a bit more easy, we will evaluate the inverse of the noise transfer function, H_{noise,R_p}. So we define

$$H_{noise,R_p}(\omega_0 + \Delta\omega) \doteq \frac{1}{T_{noise,R_p}(\omega_0 + \Delta\omega)} \tag{3.18}$$

We approximate this function with a linearization around the center frequency. So we write :

$$H_{noise,R_p}(\omega_0 + \Delta\omega) = H_{noise,R_p}(\omega_0) + \frac{dH_{noise,R_p}}{d\omega}(\omega_0) \cdot \Delta\omega \tag{3.19}$$

The first term in this expression, $H_{noise,R_p}(\omega_0)$, is equal to $G_p - G_M$ and is thus equal to zero for $G_M = G_{M,R_p}$. The second term is equal to

$$\begin{aligned} \frac{dH_{noise,R_p}}{d\omega}(\omega_0) \cdot \Delta\omega &= \omega_0 \cdot \frac{dH_{noise,R_p}}{d\omega}(\omega_0) \cdot \left(\frac{\Delta\omega}{\omega_0}\right) \\ &= 2j \cdot \sqrt{\frac{C}{L}} \cdot \left(\frac{\Delta\omega}{\omega_0}\right) \end{aligned} \tag{3.20}$$

The transfer function from the current noise source associated with R_p to the output is thus given by

$$\begin{aligned} T^2_{noise,R_p}(\omega_0 + \Delta\omega) &\approx \left|\frac{1}{2j} \cdot \sqrt{\frac{L}{C}}\right|^2 \cdot \left(\frac{\omega_0}{\Delta\omega}\right)^2 \\ &= \frac{1}{4 \cdot (\omega_0 C)^2} \cdot \left(\frac{\omega_0}{\Delta\omega}\right)^2 \end{aligned} \tag{3.21}$$

So the noise density at frequencies very close to the center frequency will be

$$\begin{aligned} \overline{dV}^2_{out}(\omega_0 + \Delta\omega) &= T^2_{noise,R_p}(\omega_0 + \Delta\omega) \times \overline{di}^2_{Rp} \\ &\approx \frac{1}{4 \cdot (\omega_0 C)^2} \left(\frac{\omega_0}{\Delta\omega}\right)^2 \times \frac{4kT}{R_p} \cdot df \\ &= kT \cdot \frac{1}{R_p \cdot (\omega_0 C)^2} \cdot \left(\frac{\omega_0}{\Delta\omega}\right)^2 \cdot df \end{aligned} \tag{3.22}$$

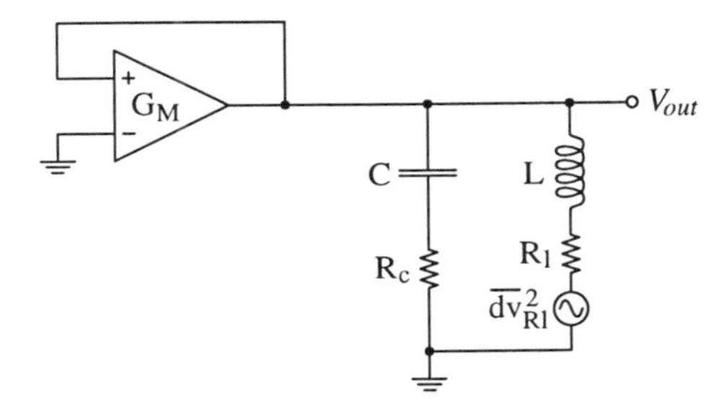

Figure 3.6. Basic oscillator with noisy R_l

Finally, we rewrite this formula with the argument enclosed in curly brackets instead of parentheses. This is just a short notation, since $\{\Delta\omega\}$ has the same meaning as $(\omega_0 + \Delta\omega)$. We also add the subscript R_p, to distinguish the several contributions to the output noise.

$$\overline{dV}^2_{out,R_p}\{\Delta\omega\} = kT \cdot \frac{1}{R_p \cdot (\omega_0 C)^2} \cdot \left(\frac{\omega_0}{\Delta\omega}\right)^2 \cdot df \tag{3.23}$$

This equation was verified by SPICE circuit simulations [HSpice96] in the frequency domain. All other contributions to the phase noise will be written in a format similar to (3.23).

The noise described here is actually amplified sideband noise. It can be split up into an amplitude modulation (AM) and a phase modulation (PM) component [Haigh 89]. If the oscillator employs an AGC circuit, the AM component will be removed for frequency offsets smaller than the AGC loop bandwidth. So the absolute value of the noise will be half of the value given by (3.23). An oscillator with a hard limiter can be regarded as having an AGC with infinite bandwidth, so the two times reduction should be valid for all frequency offsets. In practical circuits this phase noise reduction factor will be somewhere between 1 and 2. We will perform a worst-case analysis, so we will use a factor of 1 in the rest of this chapter.

3.3.2 Inductor Series Resistance R_l

Figure 3.6 shows the circuit that is analyzed in this section. Now the only parasitic resistance is R_l, which is the series resistance of the inductor. Again the loop transfer function, the resonance frequency, and the necessary transconductance are calculated in equations (3.24) to (3.26).

$$T_{loop,R_l}(s) = G_M \cdot \frac{R_l + sL}{1 + sR_lC + s^2LC} \tag{3.24}$$

$$\omega_0 = \frac{1}{\sqrt{LC}} \cdot \sqrt{1 - \frac{R_l^2C}{L}} \tag{3.25}$$

$$G_{M,R_l} = R_l \cdot \frac{C}{L} \approx R_l \cdot (\omega_0 C)^2 \tag{3.26}$$

The noise transfer function and its approximation for frequencies close to ω_0 are given in (3.27) and (3.28).

$$T^2_{noise,R_l}(s) = \frac{\overline{dV}^2_{out}}{\overline{dv}^2_{Rl}}(s) = \left[\frac{1}{1 - G_M R_l + s \cdot (G_M L - R_l C) + s^2 LC}\right]^2 \tag{3.27}$$

$$T_{noise,R_l}(\omega_0 + \Delta\omega) \approx -\frac{1}{2}\frac{1}{1 - R_l^2\frac{C}{L}}\left(\frac{\omega_0}{\Delta\omega}\right) \approx -\frac{1}{2}\left(\frac{\omega_0}{\Delta\omega}\right) \tag{3.28}$$

So the noise density at frequencies very close to the center frequency will be

$$\overline{dV}^2_{out}(\omega_0 + \Delta\omega) \approx \left|-\frac{1}{2}\right|^2 \cdot \left(\frac{\omega_0}{\Delta\omega}\right)^2 \cdot 4kT \cdot R_l \cdot df \tag{3.29}$$

which can be rewritten in a notation similar to (3.23) :

$$\overline{dV}^2_{out,R_l}\{\Delta\omega\} = kT \cdot R_l \cdot \left(\frac{\omega_0}{\Delta\omega}\right)^2 \cdot df \tag{3.30}$$

3.3.3 Capacitor Series Resistance R_c

The capacitor series resistance R_c can be analyzed in the same way as R_l. The power needed to maintain the oscillation in the presence of R_c is :

$$G_{M,R_c} = R_c \cdot \frac{C}{L} \approx R_c \cdot (\omega_0 C)^2 \tag{3.31}$$

Its noise contribution is given by

$$\overline{dV}^2_{out,R_c}\{\Delta\omega\} = kT \cdot R_c \cdot \left(\frac{\omega_0}{\Delta\omega}\right)^2 \cdot df \tag{3.32}$$

3.3.4 Effective Resistance

Now we can introduce a very important term in the evaluation of LC-tuned oscillators : the *effective resistance.* The previous three sections all rendered analog results and can be summarized with the following equations :

$$R_{eff} = R_c + R_l + \frac{1}{R_p \cdot (\omega_0 C)^2} \tag{3.33}$$

$$G_M = R_{eff} \cdot (\omega_0 C)^2 \tag{3.34}$$

$$\overline{dV}^2_{out,R}\{\Delta\omega\} = kT \cdot R_{eff} \cdot \left(\frac{\omega_0}{\Delta\omega}\right)^2 \cdot df \tag{3.35}$$

So all parasitic resistances can be reduced to one effective resistance R_{eff}, which is the equivalent of all series resistances in the loop. This resistance can be used to compare different realizations and implementations of LC-tuned oscillators, since it determines the necessary power as well as the phase noise.

It may be noticed that this effective resistance is the same as the one that would be found when substituting the parallel resistance with an equivalent series resistance, or vice versa.

The Q-factor of the LC-tank can also be expressed as a function of this effective resistance. There are three contributions to the Q-factor, one from each resistor. When combining these as done in (3.36), it is seen that the Q-factor is inversely proportional to R_{eff}

$$\begin{aligned} Q &= \frac{1}{\frac{1}{Q_{R_p}} + \frac{1}{Q_{R_l}} + \frac{1}{Q_{R_c}}} = \frac{1}{\frac{1}{R_p(\omega_0 C)} + R_l(\omega_0 C) + R_c(\omega_0 C)} \\ &= \frac{1}{R_{eff} \cdot (\omega_0 C)} \end{aligned} \tag{3.36}$$

3.3.5 Active Element G_M

The active element in the oscillator also introduces noise. It can be modeled by an output current source $\overline{di}^2_{G_M}$ equal to

$$\overline{di}^2_{G_M} = 4kT \cdot F_{G_M} \cdot G_M \cdot df \tag{3.37}$$

with G_M given by (3.34) and F_{G_M} the excess noise factor of the amplifier used. Since this noise source is the same as the one for a parallel resistance R_p, the noise transfer function is given by (3.21). So the output noise due to the active element is given by

$$\begin{aligned}\overline{dV}^2_{out,G_M}(\omega_0+\Delta\omega) &= T^2_{noise,R_p}(\omega_0+\Delta\omega)\cdot\overline{di}^2_{G_M} \\ &\approx kT\cdot\frac{1}{(\omega_0 C)^2}\cdot F_{G_M}\cdot G_M\left(\frac{\omega_0}{\Delta\omega}\right)^2\cdot df\end{aligned} \quad (3.38)$$

If we substitute the value of G_M given by (3.34) this evaluates to

$$\overline{dV}^2_{out,G_M}\{\Delta\omega\} = kT\cdot R_{eff}\cdot F_{G_M}\cdot\left(\frac{\omega_0}{\Delta\omega}\right)^2\cdot df \quad (3.39)$$

In a real circuit the transconductance used will be higher than theoretically needed. This can be due to a phase shift in the transconductor, which diminishes the real part of its transconductance value, or due to a safety margin that is usually incorporated in the design of the oscillator. We will include this fact in our equations by multiplying the noise with a factor α, representing the amount of noise the actual noisy amplifier generates in excess of an ideal noisy amplifier. To simplify the notation, we define a factor A as being equal to $\alpha\cdot F_{G_M}$. We call A the amplifier noise contribution factor. So (3.39) becomes :

$$\overline{dV}^2_{out,G_M}\{\Delta\omega\} = kT\cdot R_{eff}\cdot A\cdot\left(\frac{\omega_0}{\Delta\omega}\right)^2\cdot df \quad (3.40)$$

This equation has the same shape as the previous results, so it is easy to combine them is a general formula.

3.3.6 Conclusion

The results of the previous sections can be summarized in the following equations :

$$\overline{dV}^2_{out}\{\Delta\omega\} = kT\cdot R_{eff}\cdot[1+A]\cdot\left(\frac{\omega_0}{\Delta\omega}\right)^2\cdot df \quad (3.41)$$

$$\omega_0 = \frac{1}{\sqrt{LC}} \quad (3.42)$$

$$R_{eff} = R_l + R_c + \frac{1}{R_p\,(\omega_0 C)^2} \quad (3.43)$$

$$A = \alpha\cdot F_{G_M} \quad (3.44)$$

$$G_M = R_{eff}\cdot(\omega_0 C)^2 \quad (3.45)$$

These results were also easily verified by circuit simulations. To draw some conclusions from these calculations, we can split the phase noise up into two parts. The first one is the contribution of the resistances of the LC-tank. It is represented by the term $R_{eff} \cdot 1$. The second one is due to the active element used and is represented by the term $R_{eff} \cdot A$. It is also proportional to R_{eff} because the power needed to maintain the oscillation is determined by this effective resistance. A is usually equal to or larger than 1.

To calculate the single sided spectral phase noise density, the output noise must be integrated over a 1-Hz bandwidth and divided by the carrier power :

$$\mathcal{L}\{\Delta\omega\} = \frac{\int_{\Delta\omega-1/2}^{\Delta\omega+1/2} \overline{dV_{out}^2}\{\Delta\omega\}}{\text{carrier power}} = \frac{kT \cdot R_{eff} \cdot [1+A] \cdot \left(\frac{\omega_0}{\Delta\omega}\right)^2}{V_A^2/2} \tag{3.46}$$

This expression can be seen as the inverse of a Signal-to-Noise Ratio (SNR). A very efficient way to reduce the phase noise is enlarging the oscillation amplitude and thus making the signal larger than the noise. There is a limitation in doing this, since the maximum voltage swing at the input of the amplifier G_M is limited by the IC-technology.

The most important parameter in the design of an LC-tuned oscillator is thus the effective resistance R_{eff}. It determines the phase noise of the oscillator as well as the necessary power. To get an idea about the feasibility of low-noise oscillators, we can calculate the maximum resistance allowed to achieve the DCS-1800 spec given in (2.4). In most designs, the negative resistance is sized three times larger than theoretically necessary to ensure a reliable start-up of the oscillator. So the factor A equals three. If the oscillation amplitude is limited to 1 V, this results in :

$$R_{eff} = \frac{10^{\frac{-119\ dBc/Hz}{10}} \cdot \frac{(1\ V)^2}{2}}{0.41\ 10^{-20} \cdot [1+3] \cdot \left(\frac{1.8\ GHz}{600\ kHz}\right)^2} = 4\ \Omega \tag{3.47}$$

The power needed depends on the value of C. If an inductor of 3 nH is used, the capacitance value is 2.6 pF. This yields a transconductance of 3.5 mS. Equations (3.41) through (3.45) thus clearly illustrate the need for an inductor with very low series resistance.

3.4 PHASE NOISE OF A BASIC OSCILLATOR WITH ACTIVE INDUCTOR

Active inductors were discussed in section 3.2.2. As shown in figure 3.3(c), two transconductors g_{mi} and g_{mv} transform the impedance of a capacitor C_L into the impedance of an inductor. The inductance value is given by $\frac{C_L}{g_{mi} \cdot g_{mv}}$. So the oscillation frequency is given by

$$\omega_0 = \frac{1}{\sqrt{LC}} = \sqrt{\frac{g_{mi} \cdot g_{mv}}{C \cdot C_L}} \tag{3.48}$$

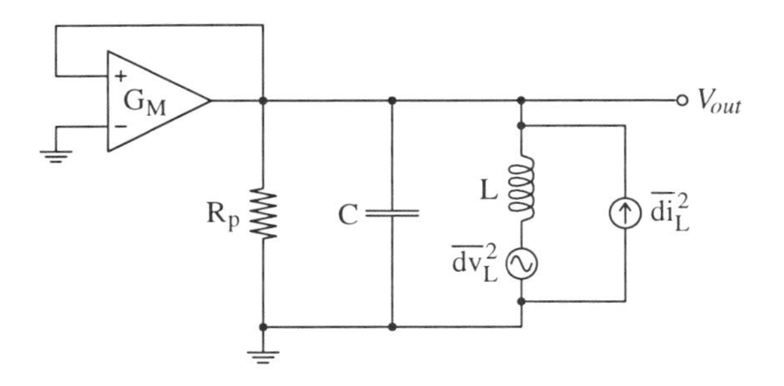

Figure 3.7. Basic oscillator with noisy active inductor

Noise originating from the active elements can be modeled by a current noise source $\overline{di}^2_L$ and a voltage noise source $\overline{dv}^2_L$. To investigate the effect of these two noise sources associated with the active inductor, figure 3.7 will be used. Only the parallel resistor R_p is included in the calculations, but the extension to all parasitic resistors can easily be made by substituting R_p with $\frac{1}{R_{eff}(\omega C)^2}$.

3.4.1 Inductor Current Noise Source

The current noise source has a value given by (3.11). The loop transfer function, the oscillation frequency and the noise transfer function are the same as calculated in sections 3.3.1 and 3.3.5. So the output noise at frequencies close to the carrier is given by

$$\begin{aligned} \overline{dV}^2_{out,Li}(\omega_0 + \Delta\omega) &= T^2_{noise,R_p}(\omega_0 + \Delta\omega) \cdot \overline{di}^2_L \\ &\approx kT \cdot \frac{1}{(\omega_0 C)^2} \cdot F_{Li} \cdot g_{mi} \cdot \left(\frac{\omega_0}{\Delta\omega}\right)^2 \cdot df \end{aligned} \tag{3.49}$$

3.4.2 Inductor Voltage Noise Source

The voltage noise source value is also given in (3.11). In this case, the loop transfer function, the oscillation frequency and the noise transfer function are the same as calculated in section 3.3.2. So the output noise at frequencies close to the carrier is given by

$$\begin{aligned} \overline{dV}^2_{out,Lv}(\omega_0 + \Delta\omega) &= T^2_{noise,R_l}(\omega_0 + \Delta\omega) \cdot \overline{dv}^2_L \\ &\approx kT \cdot \frac{F_{Lv}}{g_{mv}} \cdot \left(\frac{\omega_0}{\Delta\omega}\right)^2 \cdot df \end{aligned} \tag{3.50}$$

3.4.3 Total Noise

The total noise is calculated by combining equations (3.41), (3.49) and (3.50).

$$\overline{dV^2_{out}}\{\Delta\omega\} = kT \cdot \left[R_{eff}(1 + A) + \frac{1}{\omega_0 C} \underbrace{\left(\frac{F_{Li} \cdot g_{mi}}{\omega_0 C} + \frac{F_{Lv}}{g_{mv}}(\omega_0 C) \right)}_{M} \right] \left(\frac{\omega_0}{\Delta\omega} \right)^2 \cdot df \tag{3.51}$$

To evaluate this formula, we must now find an expression for the factor M. This factor represents the noise contribution of the active inductor. This can easily be done by substituting the expression for ω_0 given by (3.48). The result is

$$\begin{aligned} M &= \frac{F_{Li} \cdot g_{mi}}{\omega_0 C} + \frac{F_{Lv} \cdot (\omega_0 C)}{g_{mv}} \\ &= F_{Li} \cdot \sqrt{\frac{g_{mi}}{g_{mv}}} \cdot \sqrt{\frac{C_L}{C}} + F_{Lv} \cdot \sqrt{\frac{g_{mi}}{g_{mv}}} \cdot \sqrt{\frac{C}{C_L}} \\ &= \sqrt{\frac{g_{mi}}{g_{mv}}} \cdot \left[F_{Li} \sqrt{\frac{C_L}{C}} + F_{Lv} \sqrt{\frac{C}{C_L}} \right] \end{aligned} \tag{3.52}$$

Since noise has to be minimized, we need to search the conditions for minimal M. It seems this can be done by making g_{mv} much larger than g_{mi}. But in that case the voltage swing across the capacitor C_L becomes much larger than the voltage swing across C. Supposing we have already maximized the input voltage of the amplifier G_M up to the limits given by the technology, the voltage across C_L cannot be made higher. So g_{mv} has to be chosen equal to g_{mi}.

To analyze the second factor in the expression for M, we assume F_{Li} and F_{Lv} are both equal to F_L. This is not always exactly true, but deviations will be small in most practical active inductors. The conditions for minimum M are than found to be $C = C_L$, in which case M is equal to $2 \cdot F_L$, which is sometimes simplified to $M = 2$, assuming $F_L = 1$.

One last thing that can be done to clarify (3.51), is incorporating the formula for the Q-factor given by (3.36). All results can now be summarized in the following equations :

$$\overline{dV^2_{out}}\{\Delta\omega\} = kT \cdot R_{eff} \cdot [1 + A + M \cdot Q] \cdot \left(\frac{\omega_0}{\Delta\omega} \right)^2 \cdot df \tag{3.53}$$

$$\omega_0 = \frac{1}{\sqrt{LC}} = \sqrt{\frac{g_{mi} \cdot g_{mv}}{C \cdot C_L}} \tag{3.54}$$

$$R_{eff} = R_l + R_c + \frac{1}{R_p\,(\omega_0 C)^2} \tag{3.55}$$

$$A = \alpha \cdot F_{G_M} \tag{3.56}$$

$$Q = \frac{1}{R_{eff} \cdot (\omega_0 C)} \tag{3.57}$$

$$M = \sqrt{\frac{g_{mi}}{g_{mv}}} \cdot \left[F_{Li}\sqrt{\frac{C_L}{C}} + F_{Lv}\sqrt{\frac{C}{C_L}} \right] \tag{3.58}$$

$$G_M = R_{eff} \cdot (\omega_0 C)^2 + g_{mi} + g_{mv} \tag{3.59}$$

Assuming we have designed for the optimal conditions where C_L equals C, where g_{mi} and g_{mv} are both equal to g_{mL}, and if F_{Li} and F_{Lv} are both equal to 1, some of the above equations can be rewritten :

$$\overline{dV^2_{out}}\{\Delta\omega\} = kT \cdot R_{eff} \cdot [1 + A + 2Q] \cdot \left(\frac{\omega_0}{\Delta\omega}\right)^2 \cdot df \tag{3.60}$$

$$\omega_0 = \frac{g_{mL}}{C} \tag{3.61}$$

$$\begin{aligned} G_M &= R_{eff} \cdot (\omega_0 C)^2 + 2 \cdot g_{mL} = R_{eff} \cdot (\omega_0 C)^2 + 2 \cdot (\omega_0 C) \\ &= R_{eff} \cdot (\omega_0 C)^2 \cdot [1 + 2Q] \end{aligned} \tag{3.62}$$

3.4.4 Conclusion

Equations (3.53)-(3.62) allow us to evaluate LC-tuned oscillators using active inductors. The dominating term in the noise expression is the term $2Q$. So at first sight we must create a *low-Q* LC-tank. This is true, but this low Q-factor may not be achieved by increasing R_{eff}, but by increasing the capacitor C. This result doesn't surprise us, since large capacitors usually help to reduce the noise. An expected disadvantage is the proportional increase in necessary power, which is given by (3.62). So we can state that, for active LC-tanks with Q larger than one,

$$\begin{aligned} \overline{dV^2_{out}}\{\Delta\omega\} &\approx \frac{2kT}{\omega_0 C} \cdot \left(\frac{\omega_0}{\Delta\omega}\right)^2 \cdot df \\ G_M &\approx 2\omega_0 C \end{aligned} \tag{3.63}$$

We can also investigate the same numerical example as in section 3.3.6. The required capacitance value is

$$\begin{aligned} C &= \frac{\frac{2 \times 0.41\ 10^{-20}}{2\pi \times 1.8\ GHz} \cdot \left(\frac{1.8\ GHz}{600\ kHz}\right)^2}{10^{\frac{-119\ dBc/Hz}{10}} \cdot \frac{1\ V^2}{2}} \\ &= 10.4\ pF \end{aligned} \tag{3.64}$$

The transconductance associated with this capacitance is 240 mS. This is two orders of magnitude larger than the power needed for the case of passive inductors.

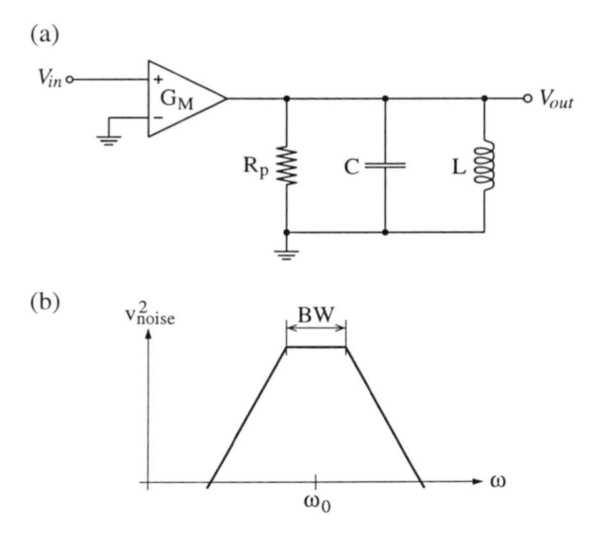

Figure 3.8. Bandpass filter : (a) Circuit; (b) Bode plot

These numbers clearly illustrate the cost of low-noise oscillators with active inductors.

3.4.5 Comparison with Bandpass Filters

Expression (3.60) can be compared with results obtained for the total integrated noise of a bandpass filter (BPF). The analogy is not a surprise, since an LC-tuned oscillator basically consists of a BPF with feedback. Figure 3.8(a) shows a general circuit diagram of a BPF.

References [Abidi CASI92] and [Crols 93] have calculated the total integrated noise at the output of the BPF. It is given by

$$v_{rms}^2 = \frac{kT}{C} \cdot [1 + A + 2Q] \tag{3.65}$$

Figure 3.8(b) shows a plot of the magnitude of the power gain vs. frequency. Since the BPF can, in first order, be regarded as having a bandwidth $BW = \frac{1}{R_p C}$, the noise density at the center frequency is

$$\overline{dV_{out}^2}(\omega_0) = \frac{v_{rms}^2}{BW} = kT \cdot R_p \cdot [1 + A + 2Q] \cdot df \tag{3.66}$$

This noise level is constant for frequencies ranging from $\omega_0 - (\frac{BW}{2})$ to $\omega_0 + (\frac{BW}{2})$. At these two corner frequencies the noise power level rolls off with a slope of 20 $dB/decade$. At a frequency offset $\Delta\omega$ from ω_0 outside the loop bandwidth the noise density is thus

$$\begin{aligned}
\overline{dV^2_{out}}(\omega_0 + \Delta\omega) &= \overline{dV^2_{out}}(\omega_0) \cdot \left(\frac{BW}{\Delta\omega}\right)^2 \cdot df \\
&= kT \cdot R_p \cdot [1 + A + 2Q] \cdot \left(\frac{1/R_pC}{\Delta\omega}\right)^2 \cdot df \\
&= kT \cdot \frac{R_p}{(R_p \cdot \omega_0 C)^2} \cdot [1 + A + 2Q] \cdot \left(\frac{\omega_0}{\Delta\omega}\right)^2 \cdot df
\end{aligned} \tag{3.67}$$

This reduces to (3.60) if R_p is replaced with its equivalent R_{eff}.

3.5 PHASE NOISE IN CRYSTAL OSCILLATORS

Crystal oscillators are commonly known for their excellent phase noise behavior. Reference [Laund UWRF92] gives a number of examples, e.g. $-137\ dBc/Hz$ at $10\ Hz$ from a $5\ MHz$ carrier. Suppose the output amplitude is $1\ V$, then the effective resistance can be calculated according to (3.41). The result is $10\ \mu\Omega$. Since this number is much smaller than the damping resistor of the crystal, another formula must apply to crystal oscillators.

Crystal oscillators can be regarded as being a special type of LC-tuned oscillators. In essence, the crystal is just a load having an infinite impedance at its parallel resonant frequency. So we can simulate a crystal oscillator by employing two capacitors and an inductor. That way we can make a completely integrated "crystal oscillator", without any external component. Figure 3.9 shows the model we will examine in this section. The basic crystal consists of C_c, C_s and L_s. Parasitic elements are modeled as follows. R_s is the combined series resistance of C_s and L_s, and R_c is the series resistance of C_c. Parallel resistance of C_c and the output impedance of the amplifier are included in R_p. If the parallel resistances of C_s and L_s must be included, they can be recalculated to an equivalent R_s.

Now the several phase noise contributions will be calculated in the same way as was done in the sections 3.3 and 3.4.

3.5.1 Parallel Resistance R_p

Figure 3.10 shows the oscillator with only R_p as a parasitic element. The loop transfer function is given by

$$T_{loop,R_p}(s) = G_M \cdot \frac{R_p \cdot (1 + s^2 C_s L_s)}{1 + sR_p \cdot (C_s + C_c) + s^2 C_s L_s + s^3 C_s C_c L_s} \tag{3.68}$$

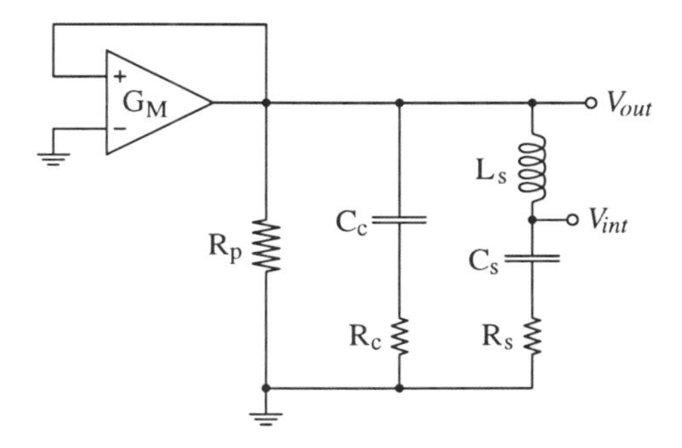

Figure 3.9. General model of a simulated crystal oscillator

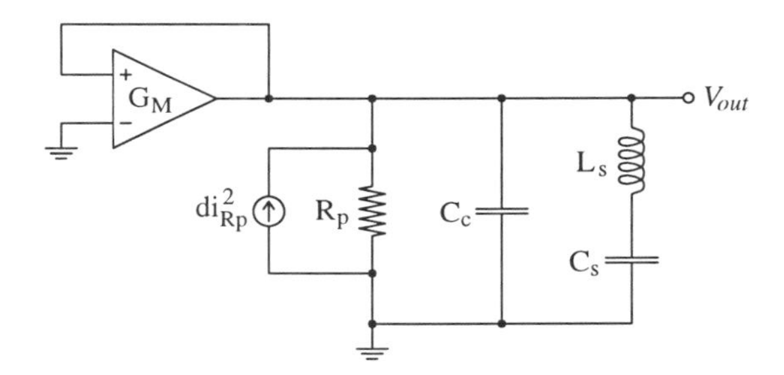

Figure 3.10. Crystal oscillator with noisy R_p

Solving the imaginary part of the loop transfer function for zero yields several solutions. They are given below :

$$\begin{cases} \omega_1 & = & 0 \\ \omega_{2,3} & = & \pm\frac{1}{\sqrt{C_s L_s}} \\ \omega_{4,5} & = & \pm\frac{1}{\sqrt{C_s L_s}} \cdot \sqrt{1+\frac{C_s}{C_c}} \end{cases} \tag{3.69}$$

The first solution is of course not useful. The next two frequencies are the series resonant frequency of the crystal. It is the frequency at which the branch consisting of C_s and L_s has zero impedance. The oscillator configuration used works only at infinite impedance, so only the last solution is useful here. It will be called ω_0 from now on.

$$\omega_0 = \frac{1}{\sqrt{C_s L_s}} \cdot \sqrt{1+\frac{C_s}{C_c}} \tag{3.70}$$

Now the transconductance G_M, necessary to have a loop transfer function exactly equal to one, can be calculated. It is given by

$$G_{M,R_p} = \frac{G_M}{T_{loop}(\omega_0)} = \frac{1}{R_p} \tag{3.71}$$

Then the transfer function from the current noise source $\overline{di}^2_{Rp}$ to the output voltage V_{out} in closed loop operation must be calculated. This results in

$$\begin{aligned} T^2_{noise,R_p}(s) &= \frac{\overline{dV}^2_{out}}{\overline{di}^2_{Rp}}(s) \\ &= \left[- \frac{1 + s^2 C_s L_s}{(G_M - G_p) \cdot (1 + s^2 C_s L_s) + s \cdot (C_s + C_c) + s^3 C_s C_c L_s} \right]^2 \end{aligned} \tag{3.72}$$

This function is approximated for small deviations $\Delta\omega$ from the center frequency ω_0, which results in

$$T_{noise,R_p}(\omega_0 + \Delta\omega) \approx \frac{1}{2} \cdot \frac{1}{\omega_0 C_c} \cdot \frac{C_s}{C_s + C_c} \cdot \frac{\omega_0}{\Delta\omega} \tag{3.73}$$

So the noise density at frequencies very close to the center frequency will be

$$\begin{aligned} \overline{dV}^2_{out,R_p}\{\Delta\omega\} &= T^2_{noise,R_p}(\omega_0 + \Delta\omega) \cdot \overline{di}^2_{Rp} \\ &\approx \frac{1}{4 \cdot (\omega_0 C_c)^2} \cdot \left(\frac{C_s}{C_s + C_c}\right)^2 \cdot \left(\frac{\omega_0}{\Delta\omega}\right)^2 \cdot \frac{4kT}{R_p} \cdot df \\ &= kT \cdot \frac{1}{R_p \cdot (\omega_0 C_c)^2} \cdot \left(\frac{C_s}{C_s + C_c}\right)^2 \cdot \left(\frac{\omega_0}{\Delta\omega}\right)^2 \cdot df \end{aligned} \tag{3.74}$$

All noise contributions can again be written in this way, as will be seen in the following sections.

3.5.2 Other Parasitic Resistors

For the other parasitic resistances in the LC-tank, i.e. R_s and R_c, similar calculations can be performed. The output noise in these cases equals :

$$\begin{aligned} \overline{dV}^2_{out,R_s}\{\Delta\omega\} &= kT \cdot R_s \cdot \left(\frac{C_s}{C_s + C_c}\right)^2 \cdot \left(\frac{\omega_0}{\Delta\omega}\right)^2 \cdot df \\ \overline{dV}^2_{out,R_c}\{\Delta\omega\} &= kT \cdot R_c \cdot \left(\frac{C_s}{C_s + C_c}\right)^2 \cdot \left(\frac{\omega_0}{\Delta\omega}\right)^2 \cdot df \end{aligned} \tag{3.75}$$

Each resistor also implies a certain amount of power to compensates for its losses. The necessary transconductances are :

$$
\begin{aligned}
G_{M,R_s} &= R_s \cdot (\omega_0 C_c)^2 \\
G_{M,R_c} &= R_c \cdot (\omega_0 C_c)^2
\end{aligned}
\tag{3.76}
$$

3.5.3 Effective Resistance and Capacitance

Equations (3.74) and (3.75) show there can be a large reduction in phase noise because of the factor $\left(\frac{C_s}{C_s+C_c}\right)^2$. As in section 3.3.4, we can define an effective resistance R_{eff} and combine the previously found results in the following equations :

$$
R_{eff} = \left[R_s + R_c + \frac{1}{R_p \cdot (\omega_0 C_c)^2}\right] \cdot \left(\frac{C_s}{C_s + C_c}\right)^2 \tag{3.77}
$$

$$
\overline{dV}^2_{out,R}\{\Delta\omega\} = kT \cdot R_{eff} \cdot \left(\frac{\omega_0}{\Delta\omega}\right)^2 \cdot df \tag{3.78}
$$

We can also define an *effective capacitance* C_{eff}, and derive a general formula for the power consumption similar to (3.34).

$$
C_{eff} = C_c \cdot \frac{C_s + C_c}{C_s} \tag{3.79}
$$

$$
G_M = R_{eff} \cdot \left(\omega_0 C_{eff}\right)^2 \tag{3.80}
$$

Using these two definitions, simple formulas for the phase noise due to the amplifier G_M and an active implementation of the inductor L_s can be derived. We can also write a simple formula for the Q-factor of the simulated crystal. It agrees with the normal expression of (3.57), but now the effective capacitance C_{eff} is used instead of C.

$$
Q_{Xtal} = \frac{1}{R_{eff} \cdot \left(\omega_0 C_{eff}\right)} \tag{3.81}
$$

So every LC-tank is completely characterized by its effective resistance and capacitance.

3.5.4 Active Element G_M

Analysis of this noise source can be done as in section 3.3.5. An output current noise source with power $\overline{di}^2_{G_M} = 4kT \cdot F_{G_M} \cdot G_M \cdot df$, with G_M given by equation 3.80, results in an output noise equal to

$$\overline{dV}^2_{out,G_M}(\omega_0 + \Delta\omega) = kT \cdot R_{eff} \cdot \alpha \cdot F_{G_M} \cdot \left(\frac{\omega_0}{\Delta\omega}\right)^2 \cdot df \tag{3.82}$$

3.5.5 Noisy Inductor L_s

We can create the LC-tank of figure 3.9 also using an active inductor. But than the noise sources of this inductor will again create output noise. Its value is given by :

$$\begin{aligned} \overline{dV}^2_{out,L}\{\Delta\omega\} &= \left[\frac{1}{2} \cdot \frac{1}{\omega_0 C_c} \cdot \frac{\omega_0}{\Delta\omega}\right]^2 \times 4kT \cdot F_{Li} \cdot g_{mi} \cdot df \\ &+ \left[\frac{1}{2} \cdot \frac{C_s}{C_s + C_c} \cdot \frac{\omega_0}{\Delta\omega}\right]^2 \times 4kT \cdot \frac{F_{Lv}}{g_{mv}} \cdot df \end{aligned} \tag{3.83}$$

We can attempt to write this formula in the general form of equation (3.51) :

$$\overline{dV}^2_{out,L}\{\Delta\omega\} = \frac{kT}{\omega_0 C_{eff}} \cdot \underbrace{\left(\frac{F_{Li} \cdot g_{mi}}{\omega_0 \frac{C_c^2}{C_{eff}}} + \frac{F_{Lv}}{g_{mv}} \cdot \left(\frac{C_s}{C_s + C_c}\right)^2 \cdot (\omega_0 C_{eff})\right)}_{M} \cdot \left(\frac{\omega_0}{\Delta\omega}\right)^2 \cdot df \tag{3.84}$$

The expression for M can be simplified to :

$$M = \sqrt{\frac{g_{mi}}{g_{mv}}} \cdot \left[F_{Li} \cdot \sqrt{\frac{C_L \cdot (C_s + C_c)}{C_s C_c}} + F_{Lv} \cdot \sqrt{\frac{C_s C_c}{C_L \cdot (C_s + C_c)}}\right] \tag{3.85}$$

Including the definition of the Q-factor of the crystal (3.81), the phase noise originating from the active inductor can be written as :

$$\overline{dV}^2_{out,L}\{\Delta\omega\} = kT \cdot R_{eff} \cdot [M \cdot Q] \cdot \left(\frac{\omega_0}{\Delta\omega}\right)^2 \cdot df \tag{3.86}$$

$$M = \sqrt{\frac{g_{mi}}{g_{mv}}} \cdot \left[F_{Li} \cdot r_C + F_{Lv} \cdot \frac{1}{r_C}\right] \tag{3.87}$$

$$r_C = \sqrt{\frac{C_L \cdot (C_s + C_c)}{C_s C_c}} \tag{3.88}$$

where we have defined r_C as a capacitor ratio.

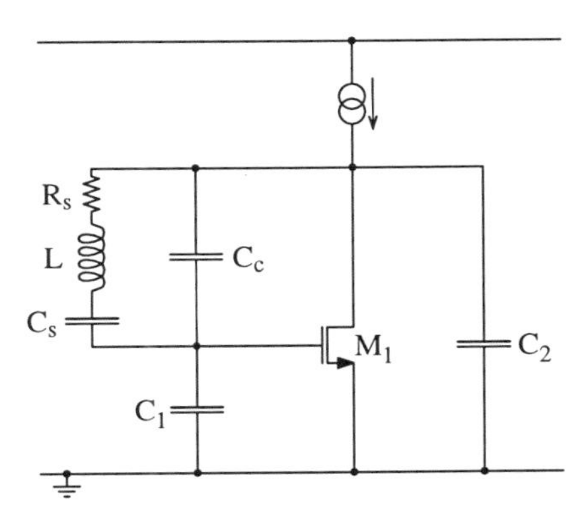

Figure 3.11. Pierce crystal oscillator circuit schematic

3.5.6 Case Study : the CMOS Pierce Crystal Oscillator

As a case study to validate the presented theory, we can examine the Pierce crystal oscillator already depicted in figure 2.18(b). The circuit is redrawn, with the crystal replaced by its electrical equivalent, in figure 3.11. In the circuits discussed in the previous sections of this chapter, the transconductance G_M providing the positive feedback required for oscillation was represented as a black box. In this case study, a single MOS transistor performs this function. The problem is of course to generate positive feedback with only a single transistor. This is not possible is a straightforward single-transistor amplifier. In the Pierce oscillator schematic two capacitors C_1 and C_2 feed a part of the oscillator signal to the source of the transistor, thereby providing a feedback with the correct sign to the oscillator crystal. The influence of only feeding back a part of the total oscillator signal to the active element will show in the noise analysis of the circuit.

The analysis of this circuit follows the general lines deducted in the previous sections. First, the oscillation frequency is calculated to be :

$$\begin{aligned} \omega_0 &\approx \sqrt{\frac{1}{L \cdot C_s} \cdot \left(1 + \frac{C_s}{C_c + \frac{C_1 C_2}{C_1 + C_2}}\right)} \\ &= \sqrt{\frac{1}{L \cdot C_s} \cdot \left(1 + \frac{C_s}{C_c + C_{12}}\right)} \\ &\quad \text{with } C_{12} = \frac{C_1 C_2}{C_1 + C_2} \end{aligned} \tag{3.89}$$

This is of course the parallel resonance frequency of the crystal, with the capacitor value C_c replaced by the sum of C_c and the series connection of C_1 and C_2. The required transconductance of the transistor M_1 needed to make the absolute value of the loop transfer function larger than one equals

$$G_M = R_s \cdot [\omega_0 \cdot (C_c + C_{12})]^2 \times \frac{(C_1 + C_2)^2}{C_1 C_2} \tag{3.90}$$

which is in perfect agreement with the previously deducted general equation (3.76), except for the factor $(C_1 + C_2)^2 / C_1 C_2$. This factor is determined by the fraction of the oscillation signal that is fed back to the active element, and how the transconductance output signal is fed into the crystal. These two elements determine how much larger the transconductance must be in order to compensate for the losses with respect to the basic feedback circuit of figure 3.9. We can immediately see that the required power in minimized for $C_1 = C_2$, which is of course why in most designs this condition is chosen. In that case the penalty for using a single transistor in the oscillator amplifier, instead of a transconductance block as in the previous sections, is a 4 times higher power consumption.

The noise transfer function from the resistor R_s to the output can, as in the previous sections, be approximated by

$$T_{noise,R_s} \approx \frac{1}{2} \cdot \frac{C_s}{C_s + C_c + C_{12}} \cdot \frac{\omega_0}{\Delta\omega} \tag{3.91}$$

which leads to an output phase noise contribution of

$$\overline{dV^2_{out,R_s}}\{\Delta\omega\} = kT \cdot R_s \cdot \left(\frac{C_s}{C_s + C_c + C_{12}}\right)^2 \cdot \left(\frac{\omega_0}{\Delta\omega}\right)^2 \cdot df \tag{3.92}$$

This is of course the same result as in equation (3.75), with C_c replaced by the equivalent $C_c + C_{12}$.

Current noise from the transistor M_1 transfers to the output according to

$$T_{noise,G_M} \approx \frac{1}{2} \cdot \frac{1}{\omega_0 \cdot (C_c + C_{12})} \cdot \frac{C_s}{C_s + C_c + C_{12}} \cdot \frac{C_1}{C_1 + C_2} \cdot \frac{\omega_0}{\Delta\omega} \tag{3.93}$$

which leads to an output phase noise contribution equal to

$$\overline{dV^2_{out,G_M}}\{\Delta\omega\} = kT \cdot R_s \cdot F_{G_M} \cdot \left(\frac{C_s}{C_s + C_c + C_{12}}\right)^2 \cdot \left(\frac{\omega_0}{\Delta\omega}\right)^2 \cdot df \tag{3.94}$$

This result is also the same as the one given by the general formula (3.82). Although the transconductance value is higher than in the previous circuit schematics, this is not reflected in the resulting phase noise. The reason for this is that the increase in transconductance is annihilated by a decrease in the transfer characteristic from the transistor to the output. It is a result that could have been expected, as the negative resistance implied by the feedback around $M1$ is exactly equal to the positive series resistance R_s, whose losses it must cancel. So the resulting output noise of both devices is the same.

3.5.7 Conclusion

We can summarize the results of the sections 3.5.1 through 3.5.5 in the following equations :

$$\overline{dV}^2_{out}\{\Delta\omega\} = kT \cdot R_{eff} \cdot [1 + A + M \cdot Q] \cdot \left(\frac{\omega_0}{\Delta\omega}\right)^2 \cdot df \tag{3.95}$$

$$\omega_0 = \frac{1}{\sqrt{C_s L}} \cdot \sqrt{\frac{C_s + C_c}{C_c}} = \sqrt{\frac{g_{mi} \cdot g_{mv} \cdot (C_s + C_c)}{C_s C_c C_L}} \tag{3.96}$$

$$R_{eff} = \left[R_s + R_c + \frac{1}{R_p \cdot (\omega_0 C_c)^2}\right] \cdot \left(\frac{C_s}{C_s + C_c}\right)^2 \tag{3.97}$$

$$C_{eff} = C_c \cdot \frac{C_s + C_c}{C_s} \tag{3.98}$$

$$A = \alpha \cdot F_{G_M} \tag{3.99}$$

$$Q = \frac{1}{R_{eff}(\omega_0 C_{eff})} \tag{3.100}$$

$$M = \sqrt{\frac{g_{mi}}{g_{mv}}} \cdot \left[F_{Li} \cdot r_C + F_{Lv} \cdot \frac{1}{r_C}\right] \tag{3.101}$$

$$r_C = \sqrt{\frac{C_L \cdot (C_s + C_c)}{C_s C_c}} \tag{3.102}$$

$$G_M = R_{eff} \cdot \left(\omega_0 C_{eff}\right)^2 + g_{mi} + g_{mv} \tag{3.103}$$

Compared to the standard LC-tank used in sections 3.3 and 3.4, there is an improvement in phase noise equal to $\left(\frac{C_s}{C_s+C_c}\right)^2$. For a real crystal, the ratio between C_c and C_s can be in the order of 200, so a 46-dB reduction of the phase noise can be achieved. That explains the good phase noise characteristics of crystal oscillators.

There is an easy way to see where this improvement comes from. Referring to figure 3.9, and neglecting the series resistance R_s, the voltage amplitude at the internal node (i.e. between L_s and C_s) equals

$$V_{int} = V_{out} \cdot \frac{\frac{1}{j\omega C_s}}{\frac{1}{j\omega C_s} + j\omega L} = V_{out} \cdot \frac{1}{1 - \omega^2 L C_s} \tag{3.104}$$

At resonance frequency ω_0 this simplifies to

$$V_{int} = -V_{out} \cdot \frac{C_c}{C_s} \tag{3.105}$$

So the signal at the internal node has a very high amplitude, which results in a large signal-to-noise ratio and thus low phase noise. That is the trick used in crystal

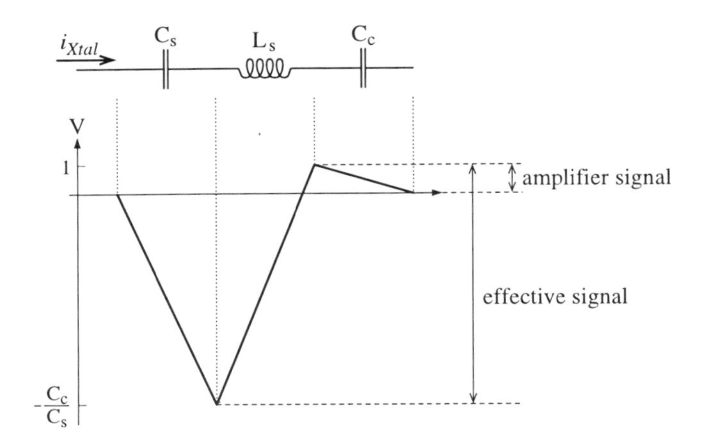

Figure 3.12. Internal voltages in a crystal

oscillators : at an internal node in the equivalent circuit the signal is very large (in the order of kilovolts), so the SNR can also be high. This high voltage does not harm the amplifier, because this one only senses the external voltage. And the crystal is only *modeled* by this LC-tank, so there is actually nowhere in the circuit such a large voltage present.

Figure 3.12 clarifies this principle. Suppose a current i_{Xtal} flows in the loop of the crystal's LC-tank. The capacitor C_s has a large negative reactance, so afterwards the voltage is very low. The inductor L_s has a positive reactance that is slightly bigger than the reactance of C_s, so the resulting voltage is positive. Finally there is the small negative reactance of C_c. The sum of all the voltages is zero, since at parallel resonance the reactances of the two branches are of opposite sign and equal value.

Basically, it is the sum of all falling (or rising) voltages that represents the factor in phase noise that was gained by the use of crystal instead of a standard LC-tank. We can call this total voltage the *effective signal* of the oscillator. For a crystal the effective signal equals

$$\begin{aligned} V_{eff} &= V_{across\ C_c} + V_{across\ C_s} = \frac{C_c}{C_s} \cdot V_{out} + 1 \cdot V_{out} \\ &= \frac{C_c + C_s}{C_s} \cdot V_{out} \end{aligned} \tag{3.106}$$

So the power of the effective signal is indeed a factor $\left(\frac{C_s}{C_s+C_c}\right)^2$ larger than the power of the output signal. The spectral purity of the oscillator output is determined by the ratio of the noise power inside the loop to the effective signal power.

We can substitute the standard LC-tank of our monolithic oscillator with a crystal-like LC-tank. When using a passive inductor, C_c must be made much larger than C_s, in order to benefit from the large internal signal. We will also assume that the effect of the parallel resistance R_p is negligible compared to the effect from the two series resistances R_s and R_c. Then the formulas applying to a crystal-like oscillator with a passive inductor are the following :

$$\begin{aligned} \overline{dV^2_{out}}\{\Delta\omega\} &\approx kT \cdot (R_s + R_c) \cdot \left(\frac{C_s}{C_c}\right)^2 \cdot [1 + A] \cdot \left(\frac{\omega_0}{\Delta\omega}\right)^2 \cdot df \\ G_M &\approx (R_s + R_c) \cdot (\omega_0 C_c)^2 \end{aligned} \tag{3.107}$$

So the power-noise product is constant : power increases quadratically with C_c, and noise decreases quadratically with C_c. The oscillation frequency is mainly determined by C_s. Although the voltages are large, these LC-tanks can be used in IC-technology. The small series capacitance can be made of metal/metal, and can withstand e.g. up to 100 V. The large capacitor only sees a small voltage, so the choice between metal/metal, poly/poly or poly/N+ is free, only depending on the area needed.

A remarkable conclusion is that, for a given inductance, the higher the required oscillation frequency, the easier it is to make the oscillator. For constant L_s and constant C_c/C_s ratio, the effective capacitance decreases quadratically with the oscillation frequency. Assuming the series resistance is mainly determined by R_s, the power needed *decreases* quadratically with ω_0.

When using an active inductor, the voltage swing at the internal node has to be limited. If the output voltage of the oscillator is already maximum, the C_c/C_s ratio cannot be larger than one. So we will make $C_c = C_s = 2C$. So each capacitor has to be twice the size of the capacitor used in an oscillator with a standard LC-tank. The effective capacitance C_{eff} is then equal to $4C$.

To minimize the phase noise we must minimize the factor M in (3.95). For the same reasons as for a standard LC-tanks, we must make g_{mi} and g_{mv} equal to each other. Assume F_{Li} and F_{Lv} are both equal to F_L, than M is minimal for $r_C = 1$. So the optimum value for C_L is

$$C_L = \frac{C_s C_c}{C_s + C_c} = C \tag{3.108}$$

The power needed for the inductor is then equal to

$$G_{M,L} = g_{mi} + g_{mv} = 2\omega_0 C \tag{3.109}$$

For high-Q inductors the phase noise and the power consumption are mainly determined by the active inductor. So the following formulae apply :

$$\begin{aligned} \overline{dV^2_{out}}\{\Delta\omega\} &\approx \frac{kT}{2\omega_0 C} \cdot \left(\frac{\omega_0}{\Delta\omega}\right)^2 \cdot df \\ G_M &\approx 2\omega_0 C \end{aligned} \tag{3.110}$$

Compared to the approximate equation (3.63) for a standard LC-tank, the phase noise is reduced by a factor of 4 (= 6 dB). The power consumption has not increased. We can again calculate the capacitance required to achieve a phase noise of $-119\ dBc/Hz$ at 600 kHz offset from a 1.8 GHz carrier with oscillation amplitude of 1 V :

$$C = \frac{\frac{0.41\ 10^{-20}}{2 \times 2\pi \times 1.8\ GHz} \cdot \left(\frac{1.8\ GHz}{600\ kHz}\right)^2}{10^{\frac{-119\ dBc/Hz}{10}} \cdot \frac{1\ V^2}{2}} = 2.6\ pF \tag{3.111}$$

This value is 4 times lower than for standard LC-tanks. But since two capacitors of twice this value are needed, plus the capacitor for the active inductor, total capacitance is 7.8 pF. So the total capacitance is now 62.5 % of the two large capacitors needed for the standard active LC-tank. The transconductance needed now is 60 mS.

So for active inductors the crystal-like LC-tank allows a reduction in power consumption by a factor of 4 with respect to a standard tank for the same noise performance. One has to keep in mind, however, that the active inductor now has to be a floating one, instead of the grounded inductor that was used in the standard LC-tank.

3.6 ENHANCED LC-TANKS

Now that we have found a way to enhance the phase noise by using special LC-tanks, we can derive other interesting structures. Referring to figure 3.12, efficient ways to generate very large *effective signals* are investigated. What matters is not the difference between the most positive voltage and the most negative voltage, but the sum of all rising (or falling) voltages. The crystal-like LC-tank was a first example.

One big disadvantage of using a large C_c/C_s ratio is the high voltages that are generated. They do not harm the operation of the circuit directly, since there is no electrical connection. But capacitive coupling might disturb other circuits on the IC. Also the large magnetic field associated with the inductors might not be negligible.

An LC-tank that does not have such large voltage swings is shown in figure 3.13(a). The internal voltages are shown in figure 3.13(b). So all nodes in the circuit have the same signal amplitude. That is why we call this tank a low-voltage enhanced LC-tank. The effective signal is now n times the external voltage swing. The area needed is n times the area of a standard LC-tank.

The case of passive inductors will be described first. Suppose the series connection of one inductor and one capacitor has resistance R. Then the following formulae

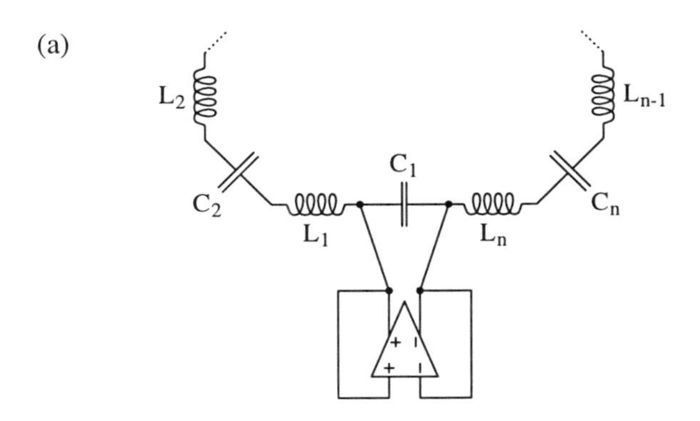

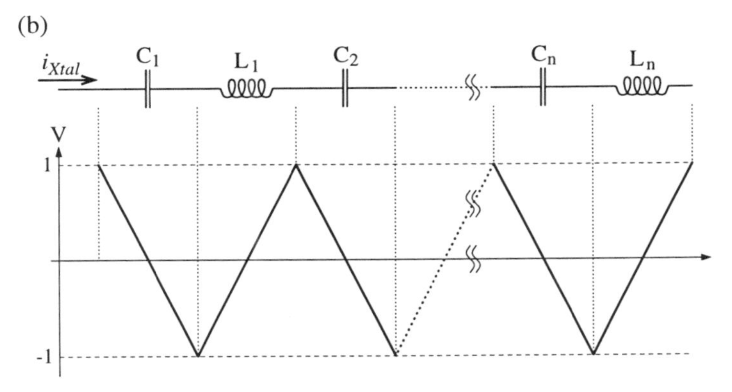

Figure 3.13. (a) Low-voltage enhanced LC-tank with n inductors and n capacitors; (b) Internal voltages

apply :

$$
\begin{aligned}
L_s &= n \cdot L \\
C_s &= \frac{C}{n-1} \\
C_c &= C \\
\omega_0 &= \frac{1}{\sqrt{LC}} \\
C_{eff} &= n \cdot C \\
R_{eff} &= n \cdot R \cdot \left(\frac{1}{n}\right)^2 = \frac{R}{n}
\end{aligned}
\tag{3.112}
$$

This leads to the following expressions for the phase noise and power consumption :

$$\begin{aligned} \overline{dV^2_{out}}\{\Delta\omega\} &\approx \frac{kT \cdot R}{n} \cdot [1 + A] \cdot \left(\frac{\omega_0}{\Delta\omega}\right)^2 \cdot df \\ G_M &\approx n \cdot R \cdot (\omega_0 C)^2 \end{aligned} \tag{3.113}$$

This shows the perfect trade-off that can be made : phase noise decreases proportionally with n, power and area increase proportionally with n. A similar trade-off can be made for active inductors. Here the formulas for phase noise, power and area are :

$$\begin{aligned} \overline{dV^2_{out}}\{\Delta\omega\} &= \frac{2kT}{n \cdot \omega_0 C} \cdot M \cdot \left(\frac{\omega_0}{\Delta\omega}\right)^2 \cdot df \\ G_M &= 2n \cdot \omega_0 C \end{aligned} \tag{3.114}$$

So phase noise decreases proportionally with n, and power and area increase proportionally with n.

3.7 OTHER PHASE NOISE SOURCES

As already indicated in chapter 2, a typical VCO noise output spectrum has three distinguishable regions. They are shown in figure 3.14, and are also covered by the semi-empirical model proposed in [Baghd IEEE65, Cutle IEEE66, Leeso IEEE66] :

$$\mathcal{L}\{\Delta\omega\} = 10 \cdot \log\left[\frac{2FkT}{P_s} \cdot \left[1 + \left(\frac{\omega_0}{2Q_L\Delta\omega}\right)^2\right] \cdot \left(1 + \frac{\Delta\omega_{1/f^3}}{\Delta\omega}\right)\right] \tag{3.115}$$

where P_s is the average power dissipated in the resistive part of the LC-tank, Q_L is the loaded quality factor and $\Delta\omega_{1/f^3}$ is the frequency of the corner between the $\Delta\omega^{-3}$ and $\Delta\omega^{-2}$ regions. This is different from the $1/f$ noise corner frequency of the transistors used in the amplifier.

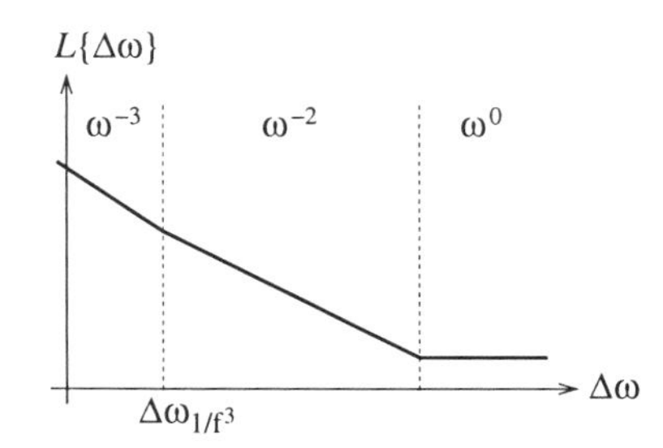

Figure 3.14. Phase noise regions in an oscillator output spectrum

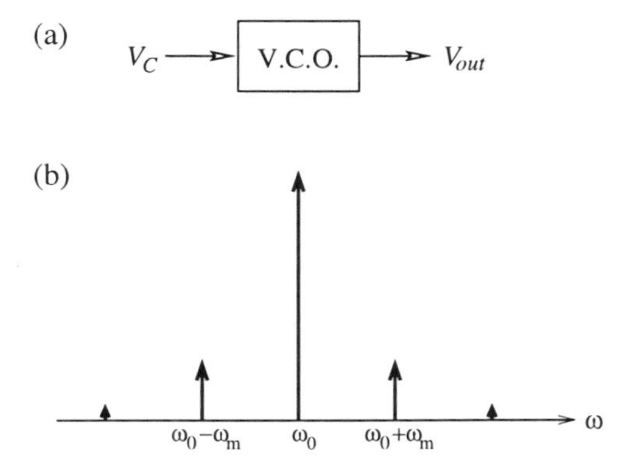

Figure 3.15. FM-modulation in a VCO

The noise sources described in all the previous sections contribute to the $\Delta\omega^{-2}$ region. They indicate the theoretical noise limit that can be achieved with a certain LC-tank. Thermal noise present at high frequencies is amplified by the positive feedback action in the oscillator to a certain level given by $\overline{dV}^2_{out}\{\Delta\omega\}$ in the various equations listed above. This section will go a little bit deeper into the phenomena that occur when the oscillator is realized with a real, non-ideal amplifier. Here, the non-linearities present in the amplifier will cause low-frequency (baseband) noise to be upconverted to the oscillator's carrier frequency, which is an extra source of phase noise.

3.7.1 FM-modulation

Most of the other phase noise sources are caused by the non-linearities present in the oscillator. Although the theory presented in the first part of this chapter assumes a linear oscillator system, this is of course not the case in any real VCO design. The transistors used have only a limited linear range, and the junction capacitances of the amplifier and those used for tuning are also not completely linear. This will cause low-frequency (baseband) noise to be upconverted to the oscillator's carrier frequency, which is an extra source of phase noise.

First the mathematical background behind the upconversion mechanism in a VCO will be discussed. A block diagram of a VCO with control input voltage V_C is shown in figure 3.15(a). This control voltage can be the actual intended tuning voltage, but can also be an internal node voltage that determines the bias of e.g. some transistor drain-bulk capacitances. Changing such an internal voltage changes the capacitance value and hence "tunes" the oscillator frequency.

The output voltage is a sine wave, the frequency of which is controlled by V_C :

$$V_{out}(t) = \cos\left[(\omega_0 + K_{vco} \cdot V_C(t)) \cdot t\right] \tag{3.116}$$

Suppose a noise signal of frequency ω_m is added to the control voltage :

$$V_C(t) = A_m \cdot \cos(\omega_m t) \tag{3.117}$$

Lengthy calculations then allow to determine the resulting output signal, which equals :

$$\begin{aligned} V_{out}(t) = \; & J_0(\beta) \cdot \cos(\omega_0 t) \\ & - J_1(\beta) \cdot [\cos((\omega_0 - \omega_m)t) - \cos((\omega_0 + \omega_m)t)] \\ & + J_2(\beta) \cdot [\cos((\omega_0 - 2\omega_m)t) + \cos((\omega_0 + 2\omega_m)t)] \\ & - \ldots \\ & \text{with } \beta = \frac{K_{vco} \cdot A_m}{\omega_m} \end{aligned} \tag{3.118}$$

with J_i the ith order Bessel function of the first kind. This is depicted in figure 3.15(b). The expression β is the modulation index. For very small β, only the first sideband is important, and $J_1(\beta)$ can be approximated by $\beta/2$. In oscillator design the amplitude A_m of the noise source is usually small, so the condition of small β is mostly fulfilled. So the phase noise at a frequency offset ω_m from the center frequency ω_0 is given by :

$$\mathcal{L}\{\omega_m\} = \left(\frac{J_1(\beta)}{J_0(\beta)}\right)^2 = \frac{\beta^2}{4} = \left(\frac{K_{VCO} \cdot A_m}{2 \cdot \omega_m}\right)^2 \tag{3.119}$$

So for a white noise source at V_C, A_m is independent of the modulating frequency ω_m and the phase noise decreases quadratically with the frequency offset. This has the same shape as the previously discussed amplified high-frequency noise, and whether or not this FM noise is dominant depends on the magnitude of A_m and K_{vco}. Usually the sensitivity of the oscillation frequency to the bias voltage of the internal nodes is low enough to guarantee this phase noise mechanism to be negligible with respect to the amplified high-frequency noise. That is why we propose the equations (3.95) through (3.103) to be the theoretical noise limit that can be achieved with a certain LC-tank.

Only the actual tuning input of the VCO has a high sensitivity, e.g. in the order of 200 MHz/V. Noise at this node can sometimes be dominant. If we take again our numerical example of an oscillator operating at 1.8 GHz, and assume a noise source with an equivalent power of 1 $k\Omega$ is present at its control input, the single-sided spectral phase noise density at 600 kHz offset is

$$\mathcal{L}\{600\,kHz\} = \left(\frac{200\,MHz/V \times \sqrt{4kT \cdot 1\,k\Omega}}{2 \times 600\,kHz}\right)^2 = 4.6 \cdot 10^{-13} = -123\,dBc/Hz \tag{3.120}$$

which is already close to the DCS-1800 spec of -119 dBc/Hz. So care has to be taken in the design of the output noise of the loop filter in the PLL, in order not to further deteriorate the synthesizer output noise.

3.7.2 $1/f$ Noise

For the $\Delta\omega^{-3}$ region, the situation is different. The shape of this curve cannot be explained by amplified high-frequency noise. It is completely due to upconverted low-frequency $1/f$ noise, as can be seen from (3.119), where A_m^2 has a $1/f$ dependency.

Modeling this with equation (3.119) is difficult, as it is not clear how the sensitivity of the oscillator center frequency to the noise source under investigation should be calculated. For example, inserting an equivalent voltage noise source at the input of the OTA of the basic LC-oscillator of figure 3.2 will probably shift the biasing point of the transistors in the amplifier. This will change some internal capacitances, such as the drain-bulk junction of the transistor, and therefore change the center frequency of the oscillator. This "sensitivity" can easily be calculated for a certain design, but the question rises if we have captured the full effect of the noise with this technique.

Studying what happens to the biasing point of the circuit still has not taken into account the non-linear effects of clipping and saturating in the amplifier. It is very likely that these non-linearities will be dominant. Calculation of this effect with the theory developed above is not possible, so other means must be found. The most straightforward, but also time-consuming method is to simulate the complete oscillator in the time domain. A properly scaled sinusoidal noise source of a low frequency ω_m can be added to the circuit, which is than simulated over a large number of periods. The spectrum of the output signal is calculated with an FFT (Fast Fourier Transform) algorithm and the relative size of the noise sidebands at $\omega_0 \pm \omega_m$ is obtained [Desme ESSC97]. Assuming the theoretical dependency on ω_m^{-3}, the complete noise spectrum at all offset frequencies can be calculated. This technique can of course also be used to simulate the phase noise in the $\Delta\omega^{-2}$ region.

3.8 STATE-OF-THE-ART INTEGRATED VCOS

Before discussing the integrated VCO designs that are presented in this work, we will take a short look at some modern state-of-the-art VCOs that were published in the open literature in the past few years. This will give us a way to evaluate the performed work with respect to the work of others in the field. Table 3.2 summarizes most data. The frequency, power, tuning range and phase noise are listed. For the phase noise, two numbers are given. In the first column, the phase noise at a certain offset from the actual center frequency as reported in the reference is given. To compare these numbers, the last column gives the value of the phase noise, recalculated to an equivalent offset frequency of 600 kHz from a 1.8-GHz carrier, assuming a dependence of 20 dB per decade on offset frequency.

Table 3.2. Comparison of integrated oscillators

Reference	*Technology [-]*	*Freq. [GHz]*	*Power [mW]*	*Tuning [%]*	*Phase noise [dBc/Hz] reported*	*equiv.*‡	*Remarks*
Relaxation oscillators							
[Banu JSSC88]	0.75-um CMOS	0.56	50	100	-90 @ 500kHz	-81	Tuning from 100kHz to 1GHz
[Sneep JSSC90]	3-GHz Bip	0.1	30	100	-118 @ 1MHz	-90	Tuning from low freq. to 150MHz
[Dobos CICC94]	9-GHz Bip	0.4	?	100	-110 @ 1MHz	-92	Tuning from 800kHz to 800MHz; Fast start-up
Ring oscillators							
[Kwasn CICC95]	1.2-um CMOS	0.74	6.5	6	-89 @ 100kHz	-97	Comparison of 3 designs
[Razav JSSC96]	0.5-um CMOS	2.2	NA	NA	-94 @ 1MHz	-91	Three-stage; differential gain stage
[vdTan ISSCC97]	9-GHz BiCMOS	2.0	NA	95	-106 @ 2MHz	-96	Two-stage CCO; stacked with mixer
LC-tuned oscillators							
[Nguye JSSC92]	10-GHz Bip	1.8	70	10	-88 @ 100kHz	-104	High-ohmic substrate; tuning with 2 tanks
[Based ESSC94]	1-um CMOS	1.0	16	0	-95 @ 100kHz	-105	Wide metal turns; substrate back-etched
[Soyue JSSC96a]	12-GHz BiCMOS	2.4	50	0	-92 @ 100kHz	-110	4-level, extra thick metal; high-ohmic substrate
[Ali ISSCC96]	25-GHz Bip	0.9	10	N.A.	-101 @ 100kHz	-110	Complete PLL; planar inductors

‡ *at 600 kHz offset from a 1.8-GHz carrier*

Continued on next page...

Continued from previous page...

Reference	*Technology [-]*	*Freq. [GHz]*	*Power [mW]*	*Tuning [%]*	*Phase noise [dBc/Hz] reported*	*equiv.*	*Remarks*
LC-tuned oscillators (cont'd)							
[Rofou ISSCC96]	1-um CMOS	0.9	10..40	14	-85 @ 100kHz	-95	Front-etched inductors; quadrature signals
[Soyue JSSC96b]	0.5-um BiCMOS	4.0	12	9	-106 @ 1MHz	-109	Thick metal (2.1 μm) and field oxide (11 μm)
[Razav ISSCC97]	0.6-um CMOS	1.8	15	7	-100 @ 500kHz	-102	Linear tuning; quadrature signals
[Dauph ISSCC97]	11-GHz BiCMOS	1.5	40	10	-105 @ 100kHz	-119	Hollow rectangular coils standard process
[Janse ISSCC97]	15-GHz Bip	2.2	43	11	-99 @ 100kHz	-116	High-Q MIS capacitor and varactor
[Parke CICC97]	0.6-um CMOS	1.6	NA	12	-105 @ 200kHz	-114	Full PLL circuit; capacitor bank for extended tuning
Presented designs							
[Steya EL94]	6-GHz Bip	1.1	1	0	-75 @ 10kHz	-106	Bonding wire inductor
[Crani JSSC95]	0.7-um CMOS	1.8	24	5	-115 @ 200kHz	-124	Bonding wire inductor; enhanced LC-tank
[Crani JSSC97]	0.7-um CMOS	1.8	6	14	-116 @ 600kHz	-116	2-level metal; conductive substrate; standard CMOS
[Crani CICC97]	0.4-um CMOS	1.8	11	20	-113 @ 200kHz	-122	2-level metal; standard CMOS

Comparison of integrated oscillators (cont'd)

The relaxation oscillators mentioned have indeed the expected characteristics : their maximum operating frequency is in the order of 1 GHz, and they all achieve linear tuning over a range of several decades. Their power consumption is moderate, but the biggest disadvantage of course is the bad phase noise spec.

Some numbers on ring oscillators are also given, but again they seem not feasible for the intended operation in portable cellular communication systems. They do achieve the required frequency of 2 GHz at a low power consumption, but the measured phase noise is too high to allow any hope in improving it to the required specs. Both [Weiga ISCAS94] and [Razav JSSC96] predict a decrease in phase noise of 10 dB per decade increase in power consumption, resulting in unrealistic values to obtain e.g. the DCS-1800 spec calculated in equation (2.4).

So we will need an LC-oscillator to do the job. A lot of work has been done on these circuits in recent years, starting with the design made by Nguyen and Meyer in 1992 [Nguye JSSC92]. This work has demonstrated the feasibility of integrated planar inductors in oscillators, and is therefore an often-used reference. It also uses a special tuning technique, where the actual oscillation frequency is situated somewhere in the middle between the resonance frequencies of two different LC-tanks, depending on which tank receives most of the power from the active circuitry. The phase noise is better than ring or relaxation oscillators, but is still far off from the required spec. This is a consequence of the fact that the inductors used are not good enough, as the tuning range in inversely proportional to the quality factor. Also, the interaction between the two tanks is expected to deteriorate the phase noise.

Starting from this point, several people have tried to improve the phase noise and hence the quality of integrated inductors. Two main options have been taken. The first route uses a process that has several layers of metal interconnects (e.g. 4 or 5) to lower the series resistance of the inductor. These designs also need a high-ohmic substrate to limit the substrate losses, a requirement that is usually fulfilled for bipolar or BiCMOS processes. Some examples are [Soyue JSSC96a, Soyue JSSC96b]. But these processes are expensive, and for integration of the RF front end with the digital baseband signal processing, a standard CMOS technology is mandatory. This is the option taken by some other designers : they do a post-processing step on finished CMOS dies that etches the substrate away underneath the inductor, thereby increasing the self-resonant frequency and removing all substrate losses. Etching can be done from the front of the wafer, as in [Chang EDL93, Rofou ISSCC96], or from the back of the wafer [Based ESSC94]. More information on this is given in chapter 5. Among these planar inductor oscillators, phase noise specs vary over a wide range. But despite all these extra costs, none of the designs achieves the DCS-1800 spec. The inductors used might be usable in several other applications, but VCO design requires a carefully optimized inductor, tuned to meet the required specs.

The oscillators developed in this work have all taken a different approach. A standard digital CMOS process in available, and no tuning, trimming, or etching is allowed. A lot of effort is put in the theoretical analysis of phase noise in oscillators, as presented in this chapter. And the integrated inductors are analyzed and optimized in order to obtain the best possible result. Chapter 4 will discuss bonding wires as an alternative for planar inductors. The realized design achieves a 5-dB better phase noise than any other oscillator reported to date [Crani ISSCC95, Crani JSSC95]. Chapter 5 discusses the optimization involved for integrated planar inductors, and hence realizes in standard CMOS a phase noise as good as or better than designs using a technology with extra costs [Crani VLSI96, Crani JSSC97]. A similar optimized oscillator in a submicron CMOS technology that uses lowly-doped bulk wafers is another 6 dB better, almost as good as the bonding wire VCO [Crani CICC97]. This last design meets the specs of the DCS-1800 system, and is used as the basic building block in a prototype frequency synthesizer discussed in chapter 7.

3.9 CONCLUSIONS

In this chapter, a definition of oscillator phase noise is given, and it is shown how this effects the signal quality in a receiver or a transmitter.

A general notation for the phase noise of LC-tuned oscillators was derived, based on the calculation of transfer functions. The results were verified by SPICE simulations. Two important terms were introduced, the effective resistance R_{eff} and the effective capacitance C_{eff}. A distinction must be made between a passive and an active implementation of the inductor.

- In oscillators using passive inductors the phase noise is completely determined by the parasitic series resistances in the loop. Enlarging the capacitance has no influence on the phase noise but increases the power consumption quadratically with capacitance. So inductors as large as possible with minimum series resistance must be constructed.

- In oscillators using active inductors the general rule of decreasing noise by increasing capacitance does work. Phase noise and power are proportional to capacitance.

Several possible improvements of the standard LC-tank with one inductor and one capacitor were studied. Starting point was the intrinsic good noise behavior of crystal oscillators. These structures can be designed in the same way as standard oscillators, using the concepts of effective resistance and capacitance. Again, active and passive inductors require different analysis.

- The phase noise of oscillators with passive inductors can be reduced by using a technique similar to that inherently present in a crystal LC-tank, where the low phase noise originates from the large signal generated on the internal node of the circuit. Passive elements can withstand these large voltages, and the amplifier only sees the small output voltage. A good trade-off can be made between noise and power.

- Oscillators with active inductors cannot withstand extremely large voltages. The internal voltage is as large as the output voltage, but since it has opposite sign, the effective signal is twice as large. So for the same phase noise, the power consumption of oscillators with enhanced LC-tanks is four times lower than that of oscillators with an active inductor that employ a standard LC-tank configuration. Low-phase-noise LC-tuned oscillators using active inductors should always use a crystal-like LC-tank when floating active inductors are available.

Some other LC-tanks were studied, whose main advantage was to reduce the internal voltage swing while holding the noise constant. This should limit the coupling between the oscillator and other circuits integrated on the same die. The most general case is the LC-tank consisting of n inductors and n capacitors. Phase noise decreases proportionally with n, but power and area increase proportionally with n. So a trade-off can be made, but practical limits (on e.g. the area) will put a maximum on the value of n. Using the concepts of effective resistance and capacitance, several other kinds of LC-tanks can be developed and analyzed.

Apart from these noise mechanisms, which put a fundamental limit on the phase noise achievable with a certain LC-tank, there are also other mechanisms by which circuit noise appears as phase noise in the oscillator's spectrum. One of them is the upconversion of baseband noise up to the carrier frequency due to non-linearities in the oscillator's amplifier or tuning capacitor. The low-frequency $1/f$ noise components thus result in $\Delta\omega^{-3}$ dependency of output power on offset frequency.

4 BONDING WIRE INDUCTANCE VCOS

4.1 INTRODUCTION

From the phase noise calculations in the previous chapter, clearly the best inductor for a low-noise oscillator is a passive inductor with a series resistance as low as possible. This is an obvious problem for fully integrated VCOs since the only type of passive inductor that is commonly used in ICs is the so-called *planar inductor* : a spiral metal trace is laid out in one or more of the standard available routing levels of the technology. But these inductors are more known for their parasitics than for their behavior as an inductor. They have a rather high series resistance, a low self-resonant frequency due to capacitive coupling to the substrate, and at high frequencies they suffer from resistive losses in the underlying silicon substrate. Optimization of these spiral or planar inductors for oscillator designs will be discussed in chapter 5.

This chapter investigates the possibilities of an alternative implementation that circumvents the problems of planar inductors. We will try to use the parasitic inductance of a *bonding wire* to our advantage. The first part of this chapter discusses the use of a bonding wire inductor in great detail. First-order formulas for the expected inductance value are given, and their correctness is demonstrated by finite-element simulations and the realization of a bipolar test VCO. Next, the possible variations in inductance value due to deviations from these first-order formulas are investigated. Finally, the parasitics of a bonding wire inductor are discussed, such as series resistance, bonding pad structure, substrate losses, etc. The second part of this chapter discusses the realization of a 1.8-GHz CMOS VCO that uses bonding wire inductors to achieve a state-of-the-art phase noise spec. Section 4.3 reports the design of the enhanced LC-tank, whereas section 4.4 discusses the implementation aspects of this circuit in a 0.7-μm standard CMOS process. The measurement results of this VCO are given in section 4.5. Finally, some conclusions are drawn from this chapter.

4.2 BONDING WIRE INDUCTORS

Since the bonding process is a standard step in the fabrication of ICs, bonding wire inductors can be regarded as being a standard passive element of a silicon CMOS technology. Normally, a bonding wire connects a pad of the IC to a pin of the package. A rule of thumb for the inductance associated with this connection is 1 nH per mm length. This is a serious problem for high-frequency in- or outputs, since the combination of this inductance with the parasitic capacitance of the bonding pad and/or the package pins limits the frequency capability of the package. For high-frequency ICs, special packages have to be used that have low parasitic capacitances, and allow for very short bonding wires to limit the inductance.

The automatic bonding equipment used in wafer fabs allows bonding wires from an IC pad to a package pin, but also a bonding wire from one IC bonding pad to another. An example of such a bonding wire that is used in a commercially available IC is shown in figure 4.1 [AD7886]. In this 12-bit, 750-kHz sampling ADC the pad-to-pad bond is used to create a connection between two nodes of the circuit with an extremely low resistance.

In this chapter we will try to use this parasitic inductance to create a high-quality inductor. A 6-nH inductor needs a capacitor of only 4 pF to resonate at a frequency of 1 GHz. These are very feasible numbers, and the bonding wire inductor offers the advantage of its intrinsic low series resistance, which will allow the fabrication of high quality factors. Also the parasitic capacitance to the substrate is limited to the capacitance of two bonding pads, so the self-resonance frequency will be very high.

4.2.1 Inductance Calculation

A differential bonding wire inductor is schematically represented in figure 4.2. Some first-order formulas to calculate the inductance value are [Green TPHP74] :

$$\begin{aligned} L &= \frac{\ell}{5} \cdot \left[\ln\left(\frac{2\ell}{r}\right) - 0.75 + \frac{r}{\ell} \right] \\ M &= \frac{\ell}{5} \cdot \left[\ln\left(\frac{\ell}{d} + \sqrt{1 + \left(\frac{\ell}{d}\right)^2} \right) - \sqrt{1 + \left(\frac{d}{\ell}\right)^2} + \frac{d}{\ell} \right] \end{aligned} \tag{4.1}$$

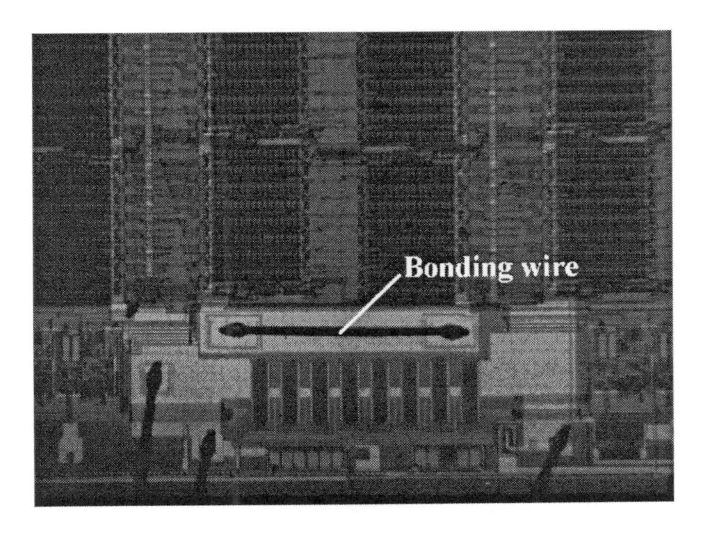

Figure 4.1. Pad-to-pad bonding wire example in a commercial IC [AD7886]

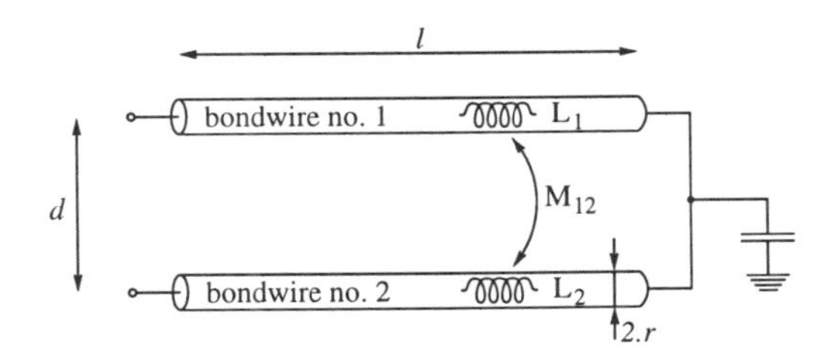

Figure 4.2. Differential bonding wire inductor

where L is the self inductance in nH of one conductor, M is the mutual inductance in nH between the two conductors, ℓ is the conductor length in mm, r is the radius of the cross section in mm, and d is the distance between the two conductors in mm. For example, two bonding wires with a standard radius of 12.5 μm and a length of 4 mm, spaced 250 μm apart, give a total inductance of 5.10 nH. These equations can be checked with finite-element simulations of bonding wire structures with a commercially available simulator MagNet [Freem 93]. For the sake of simplicity, we will first consider the 2-dimensional magnetic fields. The simulated structure is shown in figure 4.3. A circular gold bonding wire with a radius of 12.5 μm is drawn in a free air space with the shape of half a circle. Only half of a differential bonding wire inductor must be simulated because of the symmetry in the problem. In the right half shown

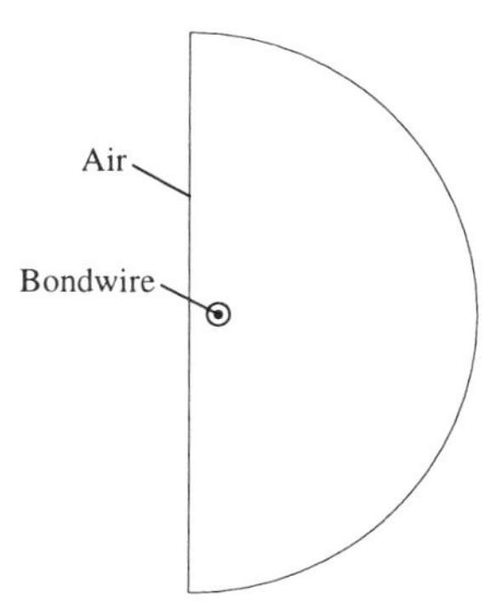

Figure 4.3. Physical structure of a bonding wire inductor for 2-D finite-element simulations

here, current flows in the bonding wire in the direction going into the page. In the other half, the same current will flow in the direction coming out of the page. Due to this symmetry, no magnetic field lines will cross the vertical symmetry axis. This is implemented by adding the appropriate constraints in the finite-element analysis. The center of the wire is at 125 μm from the left edge of the structure, which corresponds to a 250-μm spacing between the two bonding wires.

This 2-D simulation corresponds to a bonding wire inductor of infinite length, where the magnetic field effects at the beginning and the end are not important. The theoretical value for the complete differential bonding wire inductance is

$$\begin{aligned}
L_{tot} &= 2 \cdot L - 2 \cdot M \\
&= 2 \cdot \lim_{\ell \to \infty} \left\{ \tfrac{1}{\ell} \times \tfrac{\ell}{5} \times \left[\ln\left(\tfrac{2\ell}{r}\right) - 0.75 + \tfrac{r}{\ell} \right] \right\} \\
&\quad -2 \cdot \lim_{\ell \to \infty} \left\{ \tfrac{1}{\ell} \times \tfrac{\ell}{5} \times \left[\ln\left(\tfrac{\ell}{d} + \sqrt{1 + \left(\tfrac{\ell}{d}\right)^2} \right) - \sqrt{1 + \left(\tfrac{d}{\ell}\right)^2} + \tfrac{d}{\ell} \right] \right\} \\
&= \frac{2}{5} \cdot \left\{ \ln\left(\frac{d}{r}\right) + 0.25 \right\} \\
&= 0.4 \cdot \left\{ \ln\left(\frac{250\ \mu m}{12.5\ \mu m}\right) + 0.25 \right\} \\
&= 1.298\ nH/mm
\end{aligned} \tag{4.2}$$

Figure 4.4 shows a zoomed-in plot of the resulting simulated magnetic field. The grayscale level represents the flux density. The magnetic field is the largest close to the surface of the wire, and then diminishes to zero value at distances far away from the wire. The flux lines are also shown. They encircle the bonding wire. As can be seen, due to the symmetry constraint they do not cross the vertical boundary on the left

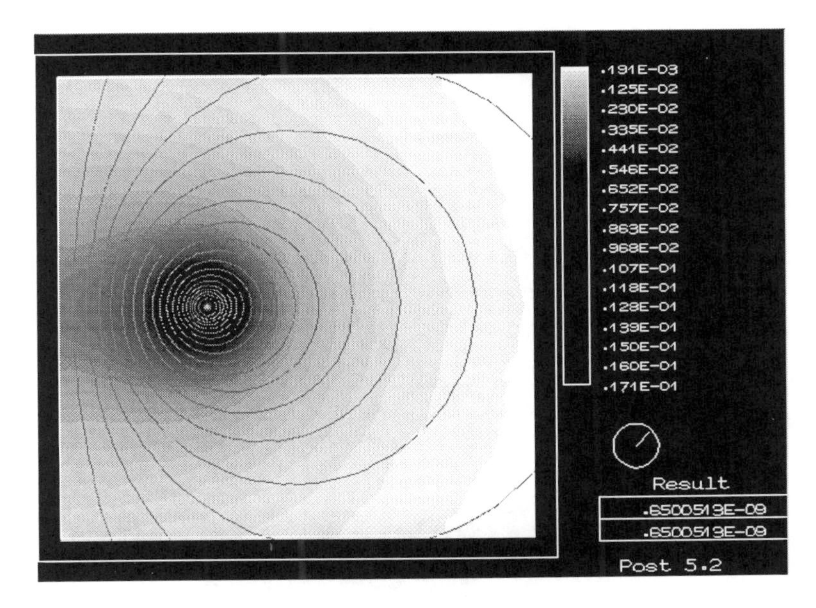

Figure 4.4. Finite-element simulation result for 2-D bonding wire inductor stucture

of the structure. In the right bottom corner, the resulting calculated inductance value is indicated. This value of 0.65005E-09 must be doubled to give the resulting total inductance value for the complete differential bonding wire inductor of 1.300 nH/mm. These results show that the theoretical formulas are extremely accurate.

For bonding wires of finite length, 3-D simulations must be performed because now the magnetic field lines do no longer flow in a 2-dimensional plane. These simulations are done in section 4.2.3.

4.2.2 Bonding Wire Test VCO

The feasibility of these bonding wire inductors is first proven in the design of a bipolar test VCO [Steya EL94]. Here, two bonding wire inductors are used in a circuit diagram as shown in figure 4.5. An IC microphotograph is shown in figure 4.6. The layout contains several oscillators. Several layout alternatives for the capacitor C are investigated, and the inductors can be bonded from chip to package or from chip to chip. The IC shown is bonded with chip-to-chip bonds.

The goal of this design is to determine whether bonding wire inductors can be used in a low noise VCO. The two inductors are not parallel to each other, so the mutual inductance can not be calculated exactly with (4.1). Each wire is 3.5 mm long, and has a cross section radius of 12.5 μm. So its self inductance is 3.9 nH. Assuming

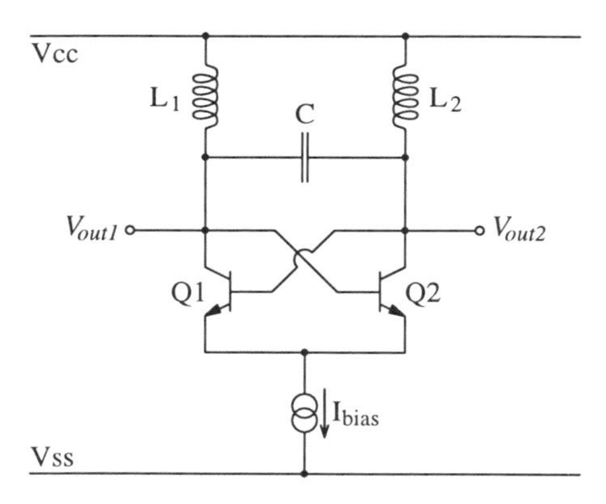

Figure 4.5. Bonding wire test VCO circuit diagram

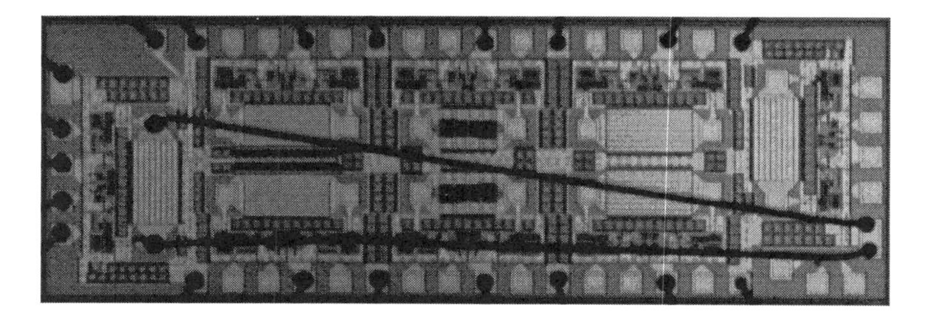

Figure 4.6. Bonding wire test VCO microphotograph

an average distance between both bonding wires of 350 μm, the mutual inductance will be approximately 1.4 nH. So the total inductance equals 5 nH, and the capacitance value C is chosen to be 4 pF. The resulting expected free-running oscillation frequency is 1.12 GHz. This can be successfully compared to a measured value of 1.13 GHz, but the exact value of the expected inductance is of course dependent on the chosen average distance between the two bonding wires.

The series resistance of the bonding wires is approximately 1 Ω, as will be calculated in section 4.2.4.2. But the quality factor of the LC-tank is also influenced by the series resistance of the capacitor C and the losses in the parasitic capacitances of the circuit and the bonding pads. These will be discussed in greater detail in section 4.2.4. In this circuit, the effective resistance R_{eff} must be estimated to be $\approx 2\ \Omega$. The single ended oscillation amplitude is -15 dBm, so the differential signal amplitude is

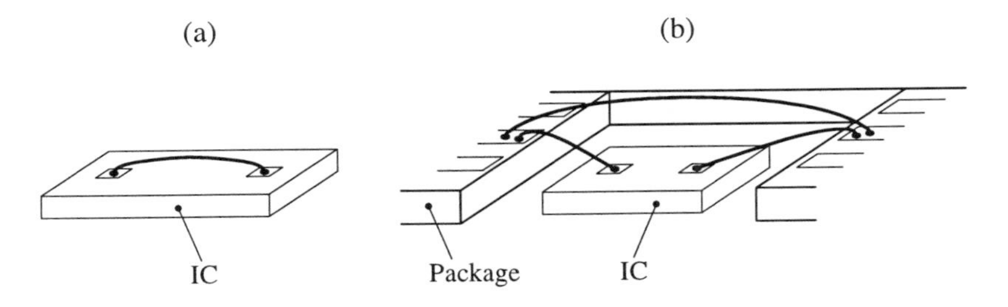

Figure 4.7. Bonding wires of fixed length : (a) chip-to-chip; (b) chip-to-package-to-chip

$2 \times 56\ mV_{pk}$. So the expected phase noise at 10 kHz offset equals

$$\mathcal{L}\{10kHz\} = \frac{kT \cdot 2\Omega \cdot [1+2] \cdot \left(\frac{1.13GHz}{10kHz}\right)^2}{(112mV)^2/2} = 5.10^{-8} = \text{-}73dBc/Hz \qquad (4.3)$$

The measured value is 2 dB better, i.e. -75 dBc/Hz, but this difference can be totally due to the inaccuracies in our calculations of the inductance value, the resistive losses, or the measurement of the oscillation amplitude.

4.2.3 *Inductance Variation*

Since we want to predict the inductance value as correctly as possible, the length of the bonding wires must be controlled accurately. This can be done using chip-to-chip bonds instead of chip-to-package bonds, as shown in figure 4.7(a). Another possible option would be to bond a wire from the die bonding pad to the IC package, then to the other side of the package, and then back to the die. This is shown in figure 4.7(b). This allows for long wires, and thus large inductance values, without the need for a large silicon die area. The sum of the lengths of the three wires is relatively independent of the position of the die in the package. But since the internal nodes at the IC pins will have a large parasitic capacitance, the range of applications of this kind of inductor will be limited.

In this section we will study the deviations on chip-to-chip bonding wire inductors, as they are the only ones that have limited parasitics to allow high quality factors. The ideal situation is the one depicted in figure 4.2. Real bonding wires, however, are never straight conductors, and the actual inductance value will differ from the one calculated with the equations (4.1). This is due to several uncertainties in the fabrication of bonding wires. Figure 4.8 shows a SEM photograph of some chip-to-chip bonding wire inductors. The gold bump on the bonding pad can be seen clearly, and from there the bonding wire starts of towards the package for the normal I/O bonding wires, or to the other end of the chip for the indicated bonding wire inductors.

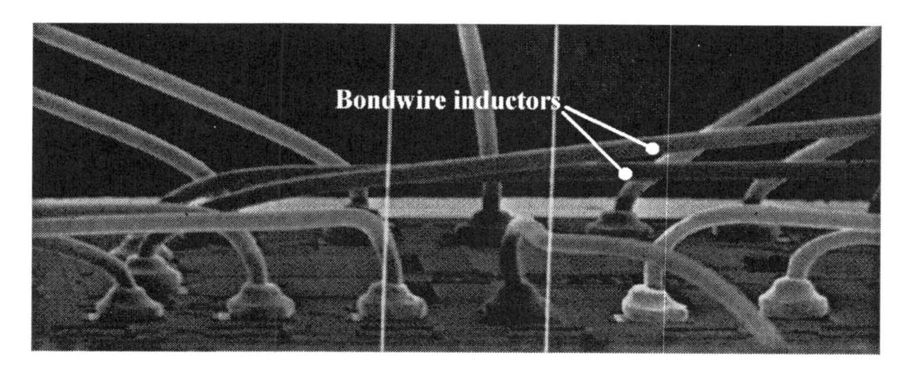

Figure 4.8. SEM microphotograph of two bonding wire inductors

The bonding wires have a certain height above the substrate as shown schematically in the top of figure 4.9. For a correct inductance calculation, the real bonding wire must be extended with a vertical wire with a length of approximately 150 μm, with a typical variation of ± 50 μm. This variation results in an uncertainty in the final inductance value. A 50-μm change in length of the vertical section of the wire corresponds with a 2-% change in total inductance value for a 4-mm long bonding wire.

Apart from that, the horizontal wire might not be completely straight. A vertical bend in the wire will cause the inductance value to change again. This bend is unpredictable, and the inductance value will therefore be different from bond to bond. This effect is checked with finite-element simulations. Starting from the 2-D structure of figure 4.3, a 3-D bonding wire is created where the position of the bonding wire varies according to a vertical bend of height h. Figure 4.9 shows the resulting simulated variation of the inductance value with the bend height. If we assume a typical vertical bend height h of 50 μm, the variation in L is less than 0.5% for bends ranging from 0 to 150 μm.

Another uncertainty is caused by a possible horizontal bend in the bonding wire. This changes the self-inductance of the wire, because of a change in the wire length, as was the case for vertical bends. The effect of this will always be small, since careful bonding can limit the deviation to e.g. 20 μm. The change in the mutual inductance between two wires is more severe. Again, finite-element simulations are used to model this effect. The results are shown in figure 4.10. For horizontal bends up to 20 μm the variation on L is limited to 4%.

The last parameter that influences the inductance value is the wire radius r. This has an effect on the self-inductance of the bonding wire. Since the radius only appears in the logarithm of (4.1), the influence will be diminished. For our example ($\ell = 4\ mm$, $d = 250\ \mu m$), a 10-% change in wire diameter corresponds to a 3-% change in inductance value.

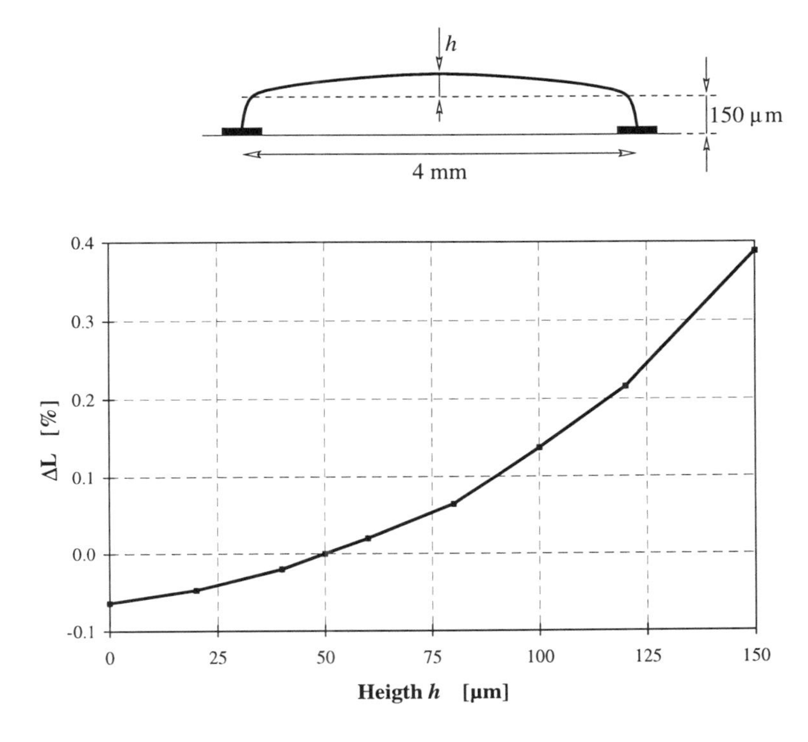

Figure 4.9. Variation of bonding wire inductance with height above the substrate

The results of all these deviations from the expected inductance value are summarized in table 4.1. Apart from the four uncertainties already discussed, a 2.5-% safety margin for imperfect modeling is incorporated. To conclude, we can safely expect the final inductance value to be within 6% of the expected value.

This is very acceptable for LC-tuned oscillators, since there will always be a larger uncertainty on the requested capacitor value. E.g. the 3σ-value of a thin oxide capacitor is typically 10% [Mietec C07]. Moreover, we intend to use metal/metal capacitors for their low series resistance. The variation on the oxide thickness between two metal layers in the order of 15%. For tunability, junction capacitors are required. The variation on their capacitance value is even worse, e.g. a MOS drain or source capacitance can vary by $\pm 20\%$. This shows that the deviation in oscillation frequency from the expected value will be mostly due to an incorrect capacitance value.

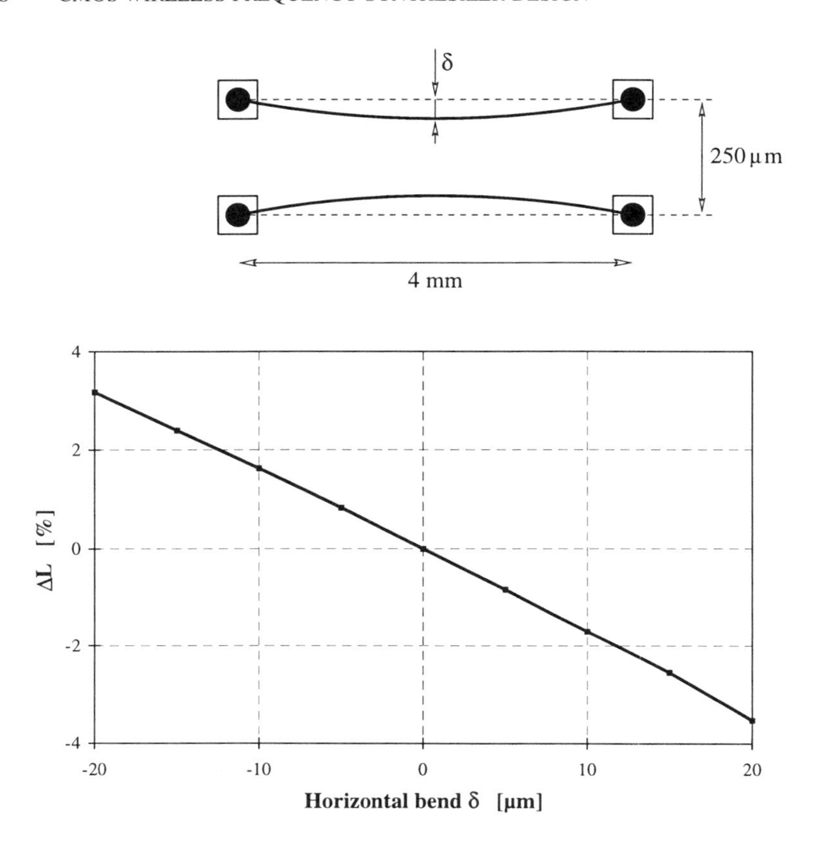

Figure 4.10. Variation of bonding wire inductance with horizontal bends

4.2.4 Parasitics

Based on the previous sections, we can conclude that bonding wires can be expected to realize good quality factors, at a reasonable inductance value. Of course, there is no such thing as an ideal inductor, and also bonding wires suffer from several parasitics. They will be discussed in the following paragraphs.

4.2.4.1 Bonding Pad Parasitics. When the inductor is used differentially, the internal node will be a common mode node, and the parasitic effect of the bonding pads at that end of the bonding wire will not be important. So the only parasitic capacitance is the one from the two bonding pads at the beginning of the bonding wires. A bonding

Description of imperfection	*Typical worst case variation*	*Effect on L*
Extra 150-μm vertical wire	$< 50\ \mu m$	2 %
Vertical bend (50 μm)	0-150 μm	0.4 %
Horizontal spacing	$< 20\ \mu m$	4 %
Wire diameter	$< 10\%$	3 %
Imperfect modeling	—	2.5 %
Total		6 %

Table 4.1. Variation in inductance value with bonding wire parameters

pad can be modeled by a capacitor in series with the resistance of the substrate. This is shown in figure 4.11.

The capacitor value C_{pad} will depend on the required bonding pad structure in the specific technology used in the design. For example, a bonding pad can sometimes consist of only a square area with size $100 \times 100\ \mu m^2$ in the highest metal layer of a double metal process. If the oxide thickness between the metal2 level and the substrate is approximately 2 μm, the bonding pad capacitance equals only

$$C_{pad} = \frac{\varepsilon_{ox} \times Area}{Thickness} = \frac{33.8\ pF/m \times (100\ \mu m)^2}{2\ \mu m} = 170\ fF \tag{4.4}$$

But e.g. in the technology used for the bonding wire oscillator described in the second part of this chapter, a bonding pad must be realized as a stack of metal2, metal1 and poly-interconnect layers. The oxide thickness from bonding pad to the substrate is then determined by the oxide thickness underneath the poly-layer, and the capacitance per area of such a bonding pad is 0.08 $fF/\mu m^2$ [Mietec C07], resulting in a total value of not less than 800 fF. This is alleviated by putting a floating N-well underneath the bonding pad, as was already shown in figure 4.11. This N-well also serves as an isolation of the bonding pad from the substrate in case the field oxide breaks down during bonding. We can assume that this N-well/P-substrate junction is biased at zero volts, since any charge causing a different biasing point will leak away through the parasitic parallel resistance of the junction. At zero volt bias, the N-well/P-substrate junction has a capacitance per area of 0.075 $fF/\mu m^2$ and a sidewall capacitance of 0.7 $fF/\mu m$ [Mietec C07], resulting in a total capacitance of approximately 1 pF. Since this junction capacitance is situated in series with the already calculated oxide

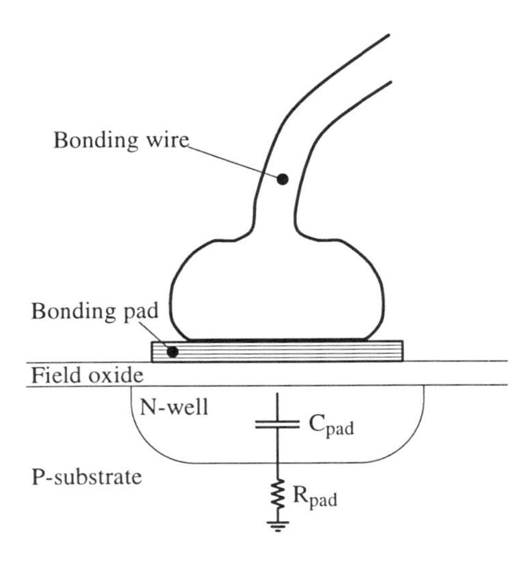

Figure 4.11. Parasitic elements of a bonding pad

capacitance, the total capacitance from bonding pad to ground can be approximated by

$$C_{pad} = \frac{1}{\frac{1}{800\ fF} + \frac{1}{1\ pF}} = 450\ fF \tag{4.5}$$

This parasitic capacitance determines the self-resonant frequency of the inductor. For a differential bonding wire inductor as in figure 4.2, the two bonding pad capacitors are placed in series, so the resulting parallel capacitance is typically 200 fF. With an inductance value of e.g. 5 nH, this yields a self-resonant frequency of 5 GHz.

Making an estimation about the value of R_{pad} is more difficult, since a wide variety of wafers can be used for processing, and the doping level of the wafer determines the resistivity of the substrate and hence R_{pad}. We will make a distinction between heavily doped and lowly doped substrates. Most submicron CMOS processes, and also the one used in the design described in the second part of this chapter, use so-called epi-wafers. A schematic cross section of such a wafer is shown in figure 4.12(a). These wafers consist of a heavily doped substrate, on top of which a layer of monolithic lowly-doped silicon, the epi-layer, is deposited. The epi-layer doping level is low to allow an optimal doping level in the N- and P-well to obtain the best transistor characteristics. The P-well is not isolated from the substrate ground. Epi-wafers are more expensive than normal wafers, but yet they are used because of the high conductivity in the substrate. This offers a lowly resistive path to ground for substrate currents that are inserted into the substrate by e.g. hot-electron effects or switching digital cir-

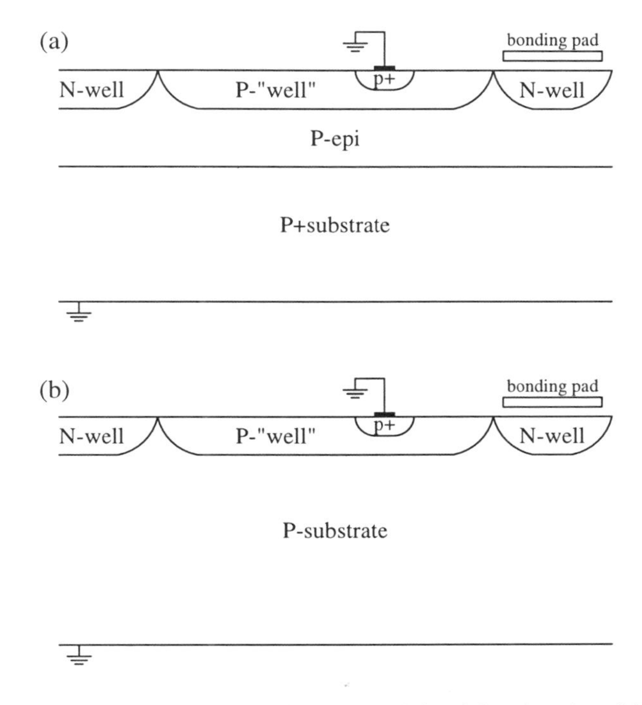

Figure 4.12. Silicon wafer structures (not to scale) : (a) epi-wafer; (b) lowly doped wafer

cuits. It also prevents problems associated with latch-up. Epi-wafers may also offer an area reduction since it is not always necessary to lay out a large number of substrate contacts to create a sufficient ground.

So the resistance R_{pad} can be split up into two parts. The first is the resistance of a small distance (typically less than 10 μm) through the epi-layer to the heavily doped substrate. The second part is the resistance through this substrate to the actual ground plane reference. To give an idea of the order of magnitude for the epi resistance, suppose the epi-layer is p-type with a doping concentration of $5.10^{14}\ cm^{-3}$. The conductivity ρ_{epi} that corresponds to this doping level is 25 Ωcm [Gray 93]. With a length of 10 μm, and a cross section area of $(100\ \mu m)^2$, this results in a resistance of

$$R_{epi} = \frac{\rho_{epi} \cdot length}{area} = \frac{25\ \Omega cm \cdot 10\ \mu m}{(100\ \mu m)^2} = 250\ \Omega \tag{4.6}$$

The second part of R_{pad} is more difficult to calculate. If the backside of the die is attached with a conductive glue to the package ground plane, there is a path from the top of the substrate to the bottom. The length of this path is typically 625 μm, depending on the wafer thickness. The cross section area at the top of the wafer is of course $(100\ \mu m)^2$, but this spreads out to a larger area since the complete back plane of the wafer is at ground potential. With a substrate doping level of $10^{19}\ cm^{-3}$, or a conductivity of $0.01\ \Omega cm$, and if we do not account for the larger cross section area at the bottom, the substrate resistance is

$$R_{subs} = \frac{\rho_{subs} \cdot length}{area} = \frac{0.01\ \Omega cm \cdot 625\ \mu m}{(100\ \mu m)^2} = 6.25\ \Omega \tag{4.7}$$

The actual value will be even smaller, so we can conclude that R_{subs} is negligible with respect to the R_{epi}. One extra remark can be made on the contact resistance of the back plane. To make a perfect contact with the package ground plane, the back side of the wafer should be polished and coated with a conductive layer. Since this is an extra cost for processing, it is not always standard procedure. If one tries to make a conductively glued contact anyway, the resulting contact resistance is not known, since at the back side of the wafer several oxides may have grown during high-temperature processing steps. If the contact resistance is too large, R_{subs} will not be determined by the vertical path to the bottom of the wafer, but the horizontal path towards substrate ground contacts at the top of the wafer.

If we consider the case where substrate contacts are placed relatively close to the bonding pad, also R_{epi} is not well known since now there will also be a horizontal resistive path in parallel with the vertical one. The actual value of this depends completely on the number and position of the substrate contacts, so no general statement can be made. The series resistance associated with the sidewall capacitance of the N-well will be much lower than that of the bottom plate, since the doping level of the P-well is higher than the doping level of the epi-layer. The N- and P-well layers usually have a sheet resistance in the order of $1\ k\Omega/\Box$. If no N-well is used underneath the bonding pad, the total resistance will be determined by the path through the more conductive P-well layer. We estimate that, if a large number of substrate contacts is placed close to the bonding pad, the resistance can be decreased to a value smaller than $100\ \Omega$.

The same thing can be said about bonding pads on lowly-doped wafers. This wafer type is shown in figure 4.12(b). Here the substrate resistance is also totally dependent on the number and the position of the substrate contacts. The only way to determine its value in a specific case is the use of finite-element simulations. Measurement is difficult because carefully calibration is required for the parasitics of the measurement probes, and this at very high frequencies. The 3-dB frequency of a 400-fF capacitance and a 100-Ω resistance is situated at 4 GHz.

The resistance R_{pad} will increase the losses in the LC-tank, but its effect can be small when a larger capacitor is placed in parallel with the bonding pad.

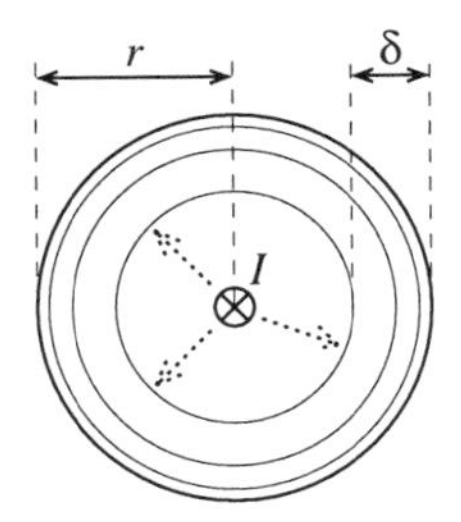

Figure 4.13. Representation of the skin effect in a bonding wire

4.2.4.2 Inductor Series Resistance. The DC series resistance of a bonding wire is given by

$$R_{wire} = \frac{\rho \cdot \ell}{A} \qquad , A = \pi \cdot r^2 \tag{4.8}$$

where ρ is the conductance of gold, which equals 2.35 $\mu\Omega cm$, ℓ is the length of the wire and A is the area of its cross section. For a bonding wire with a radius of 12.5 μm, this area equals $4.9 \cdot 10^{-4}$ mm^2. So the resistance is 0.05 Ω per mm length, which is a very low value. A differential inductor of 4 mm length has a total resistance of 0.4 Ω, which gives in combination with its inductance value of 5.1 nH a possible quality factor at 1 GHz of $(2\pi \cdot 1GHz \times 5.1nH)/0.4\Omega = 80$!

However, the skin effect will decrease the conductivity of the bonding wire and hence decrease the quality factor. Due to interactions of the high-frequency magnetic field in the conductor, the current is pushed to the outside of the conductor. This is shown schematically in figure 4.13. The center of the conductor is no longer used for current flow, and the effective resistance will increase. A skin depth δ is defined as being the equivalent thickness of a hollow conductor that has the same high-frequency resistance :

$$\delta = \sqrt{\frac{2 \cdot \rho}{\omega \cdot \mu}} \tag{4.9}$$

The permeability of gold, μ_{Au}, equals $4\pi \cdot 10^{-7}$ F/m. At a frequency of 1 GHz the skin depth of gold is equal to 2.4 μm. So only a part of the cross section conducts. The area of this part is approximately equal to $2\pi r \cdot \delta = 1.9\ 10^{-4} mm^2$, which results in a series resistance of 0.125 Ω per mm length. This explains the total resistance value of 7 $mm \times 0.125\ \Omega/mm \approx 1\ \Omega$ that was taken for the bipolar bonding wire test VCO described in section 4.2. At even higher frequencies, the skin effect dominates more and more, e.g. at 2 GHz the resistance is 0.175 Ω/mm. This certainly limits the possibilities of bonding wire inductors, so the theoretically available self resonance frequency of 5 GHz will never be used.

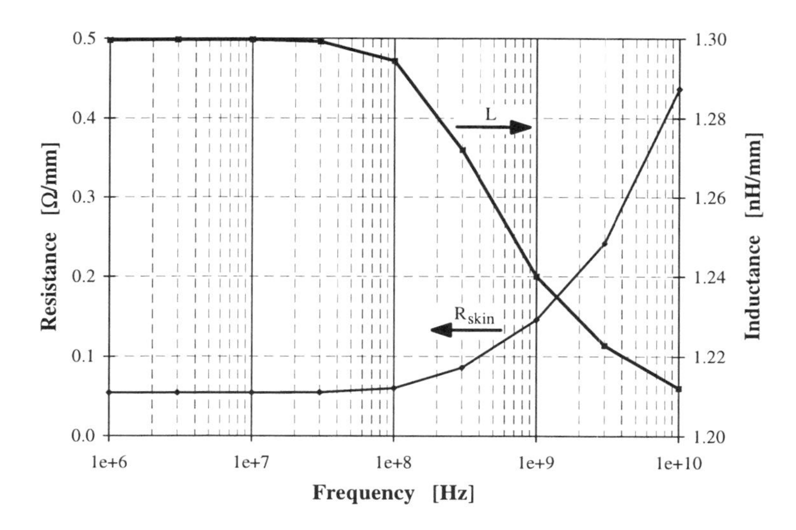

Figure 4.14. Simulation result of the skin effect in a bonding wire inductor; bonding wire radius is $12.5\ \mu m$, spacing is $250\ \mu m$

These theoretical values can also be checked for their accuracy by frequency-domain finite-element simulations on the structure of figure 4.3. Figure 4.14 shows the variation of the effective high-frequency resistance R_{skin} of a single bonding wire. So to calculate the total resistance of the full differential bonding wire inductor one must double this number. The simulated value varies from $0.054\ \Omega/mm$ at low frequencies to as much as $0.44\ \Omega/mm$ at $10\ GHz$, which is one order of magnitude larger. At $1\ GHz$, the series resistance equals $0.146\ \Omega/mm$, so the skin effect is even more severe than was expected by the first-order calculation of the skin depth δ. The variation of the inductance value is also indicated in figure 4.14. A decrease from $1.30\ nH/mm$ at low frequencies to $1.21\ nH/mm$ at $10\ GHz$ is seen. This change is also a consequence of the skin effect, which redistributes the magnetic field and thus changes the inductance. At 1-GHz frequency, the inductance value equals $1.24\ nH/mm$.

4.2.4.3 Substrate Loss. A part of the magnetic field also penetrates deep into the silicon substrate. There an induced electrical field is generated that causes a current flow in a direction such that it opposes the original magnetic field. The consequence of this is a lower total inductance value (because the resulting magnetic field is smaller) and extra resistive loss. This is again checked with finite-element simulations. The simulated structure is shown in figure 4.15(a). It is almost the same as figure 4.3, but now a conductive substrate is situated $150\ \mu m$ under the circular aluminum bonding

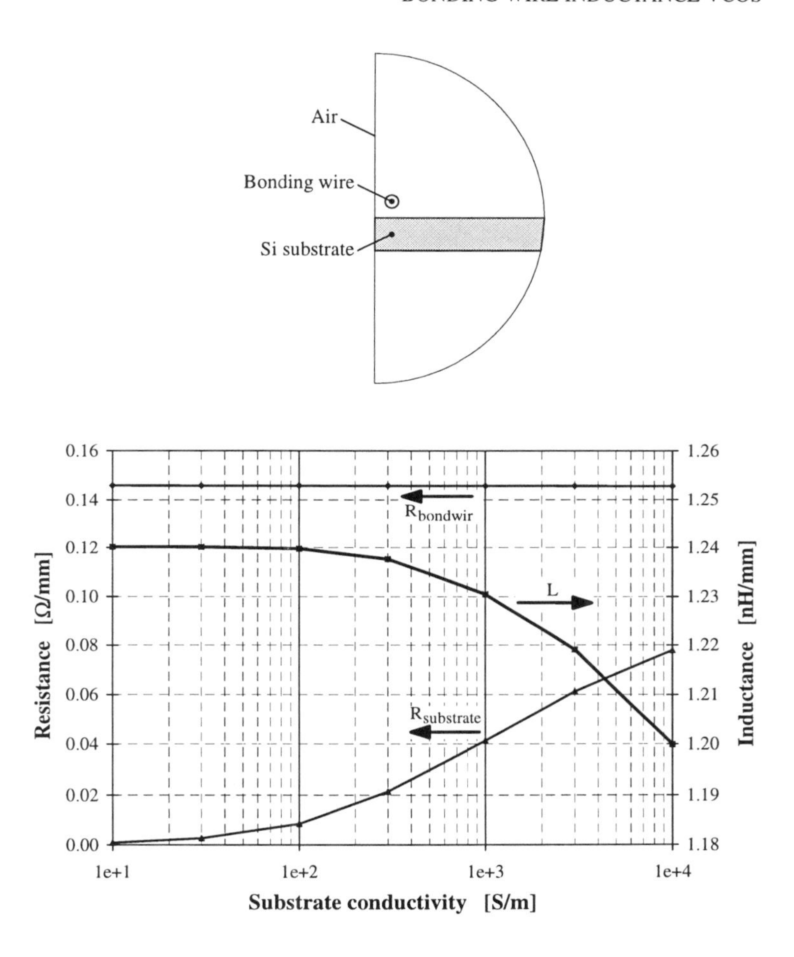

Figure 4.15. Substrate loss of a bonding wire inductor : (a) Physical structure for 2-D finite-element simulation; (b) Simulation result for a frequency of 1 GHz

wire with a radius of 12.5 μm. A current I is forced in the bonding wire in the direction going into the page, and in the substrate an induced current I_{subs} will flow in the opposite direction.

At low frequencies, the simulated inductance value is 1.300 nH/mm, which is the same as before. So the conductive substrate has no noticeable effect. However, at a frequency of 1 GHz the induced electrical field is large enough to seriously influence the inductor's performance, especially at very high substrate doping levels. Figure 4.15(b) shows the variation of the inductance and the equivalent series resistance as a function of the substrate conductivity. The simulations were done at a frequency of 1 GHz. The resistance is split up in two parts : $R_{bondwire}$ causes losses in the bonding wire itself, whereas $R_{substrate}$ is an equivalent series resistance that causes the same amount of losses as the resistive losses in the substrate. The bonding wire resistance stays constant at the in section 4.2.4.2 obtained value of 0.146 Ω/mm. The substrate loss, however, increases from a negligible value for lowly doped substrate to a value of 0.078 Ω/mm for a substrate conductivity of 10^4 S/m. This conductivity corresponds to a p-type doping level of 10^{19} cm^{-3}.

4.2.4.4 Magnetic Coupling. Since the dimensions of the differential bonding wire inductor are rather large with respect to the normal sizes of integrated circuits, there is certainly an area cost associated with the use of them. However, underneath the bonding wires a lot of free space is available that can be used to place some non-critical parts of the circuits. But the large size of the inductor also means the magnetic field will be sensed in a large part of the circuit. If a digital circuit with fast switching times is placed close to the bonding wires, the magnetic field will be disturbed by the di/dt current changes. So spurious peaks will appear in the oscillator's output spectrum at a distance from the carrier equal to the clock frequency of the digital circuit. So the area underneath the bonding wires must be reserved for non-critical analog circuits that will not disturb the inductor's magnetic field.

4.3 ENHANCED BONDING WIRE LC-TANK

Based on the low-voltage enhanced LC-tank described in section 3.6, and on the intrinsic possibilities offered by bonding wire inductors, a 1.8-GHz oscillator with an enhanced bonding wire LC-tank was developed. This section describes the design of the 4-bonding wire structure, shown in figure 4.16.

To see what actually happens in this enhanced LC-tank consisting of 2 capacitors and 2 differential bonding wire inductors, the LC-tank is shown in figure 4.16(a) and the internal voltages are drawn as in section 3.6 in figure 4.16(b). When a current i_{LC} flows in the loop of the tank, the inductors L_1 and L_2 cause a voltage drop, whereas the capacitors C_1 and C_2 have a negative reactance and cause an increase in the voltage. If the same current would flow in a standard LC-tank consisting of one inductor $L_{12} = L_1 + L_2$ and one capacitor $C_{12} = (C_1^{-1} + C_2^{-1})^{-1}$, the total voltage across the LC-tank would be twice as large. So the phase noise would be 6 dB better, since the noise level hasn't changed. So we can look at the enhanced LC-tank as having an effective signal that is twice as large as the external signal across the nodes of the LC-tank. Mathematically, the effective resistance is four times lower than the actual sum of all equivalent series resistance of the two inductors and the two capacitors. But since this sum of series resistances is twice as large for this circuit as for a tank with a single L and a single C, the overall improvement in phase noise is only 3 dB. The effective capacitance is two times larger than a single capacitor C, which yields a two times larger power consumption.

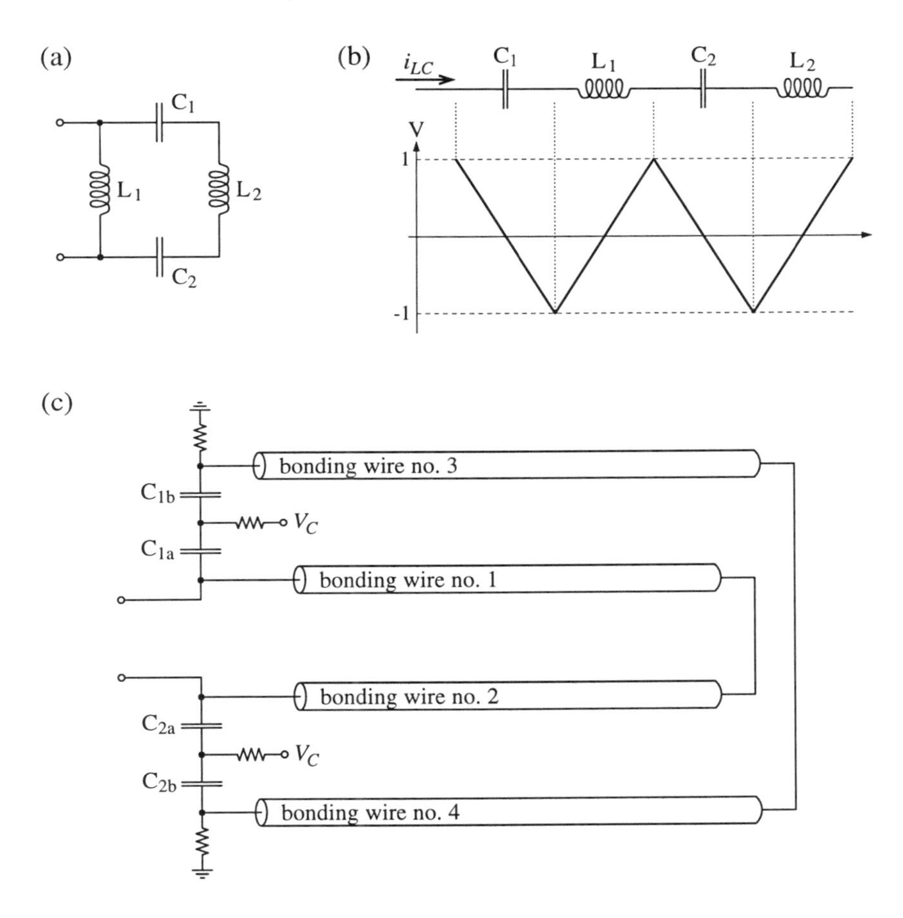

Figure 4.16. Low-voltage enhanced bonding wire LC-tank : (a) Circuit; (b) Internal voltages; (c) Physical structure

The LC-tank is implemented using the differential bonding wire inductors as shown in figure 4.16(c). Four bonding wires (two per inductor) are used. All wires are spaced 260 μm apart, and their length is 3.0 mm for the inner ones and 3.4 mm for the outer ones. Their placement next to each other causes them to interact, an effect that was accounted for by incorporating the appropriate mutual inductances in the simulations. Because of the interaction, the effect of horizontal bends in the wires will not be as severe as was calculated in section 4.2.3. If one of the mutual inductances increases because of a smaller spacing, another one will decrease.

The bonding wire sizes are determined by several factors, such as the desired oscillation frequency (1.8 GHz), maximum allowable chip area, capacitor sizes of the tank's capacitors and the parasitics of the amplifier, power consumption and of course the resulting phase noise. Making the bonding wires shorter will decrease their resistance, and hence improve the phase noise, but the smaller inductance value requires a larger capacitance and increases the power consumption. Putting the bonding wires further apart will limit the negative mutual inductance and increase the inductance value for the same resistance, but this requires a larger chip area.

This four-bonding wire structure is also simulated with the finite-element simulator MagNet [Freem 93]. Again, only half of the total structure must be modeled because of the symmetry. Figure 4.17(a) shows the two bonding wires with radius 12.5 μm, one at 130 μm from the symmetry axis with a current flowing into the page, and one at 390 μm from the symmetry axis with a current flowing out of the page. They are situated 150 μm above a 625-μm thick substrate with a resistivity of 0.01 Ωcm or a conductivity of 10^4 S/m.

The resulting magnetic field is plotted in figure 4.17(b). The grayscale level represents the flux density and the flux lines are also shown. They encircle the bonding wires, but they are not completely symmetrical around the horizontal axis. This is the influence of the conducting substrate, in which the induced currents push the flux lines away.

The skin effect causes the bonding wire's series resistance to be as high as 0.19 Ω/mm at 1.8 GHz instead of 0.05 Ω/mm at DC. The substrate losses add another equivalent series resistance of 0.11 Ω/mm to this value, resulting in a total of 0.3 Ω/mm. The current redistribution of the skin effect and the substrate currents decrease the inductance value by 9 %. This MagNet data is easily incorporated in the Spice circuit file to achieve simulation results as accurately as possible.

For frequency tuning, the capacitors were realized as the series connection of two capacitors, as already indicated in figure 4.16(c). The first one (C_{1a} and C_{2a}) is a Metal1/Metal2 capacitor ($\approx 2.7pF$) and serves as a DC voltage shift. The amplifier will be sized to have a DC output voltage that is approximately half-way the power supply. The difference between this voltage and the tuning voltage V_C stands across these capacitors. A Metal/Metal capacitor is used instead of a Poly/N+ one because of its lower series resistance. The other capacitors (C_{1b} and C_{2b}) are P+/N-well junction capacitors ($\approx 2.8pF$) which have their anode biased to ground potential. They enable the tuning of the center frequency with the control voltage V_C. This DC-biasing scheme prevents the junction capacitors from being forward biased for all possible control voltages. A schematic cross section of the capacitor structure is shown in figure 4.18. A disadvantage of the capacitor series connection is a decrease in tuning range, since the total capacitance only changes by 0.5% for a 1-% change in the junc-

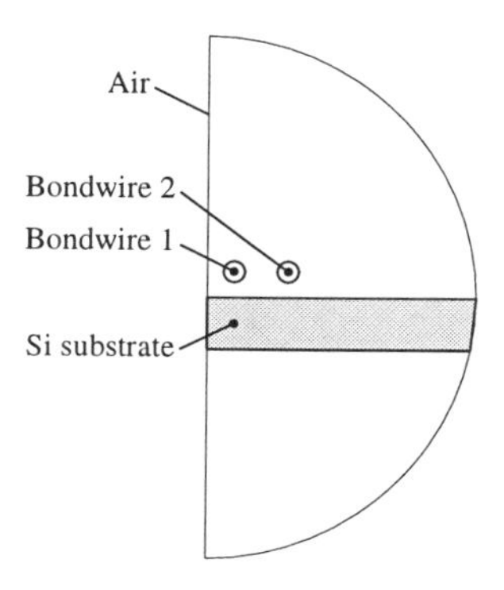

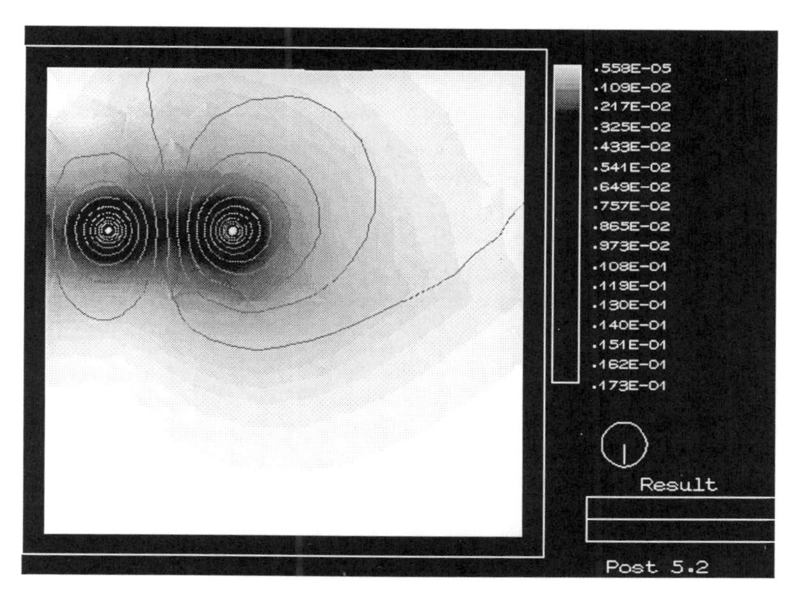

Figure 4.17. (a) Enhanced bonding wire LC-tank structure; (b) Finite-element simulation result

tion capacitance. The ratio between the values of the metal and the junction capacitor is chosen such that the voltage swing at the node in between them is minimal. This is necessary because the junction's parasitic N-well capacitance and the metal bottom plate parasitic capacitance is situated at this node.

For the simulation of the LC-tank, all the parasitics are included. The bonding wires are modeled with the appropriate self- and mutual inductances, and are given

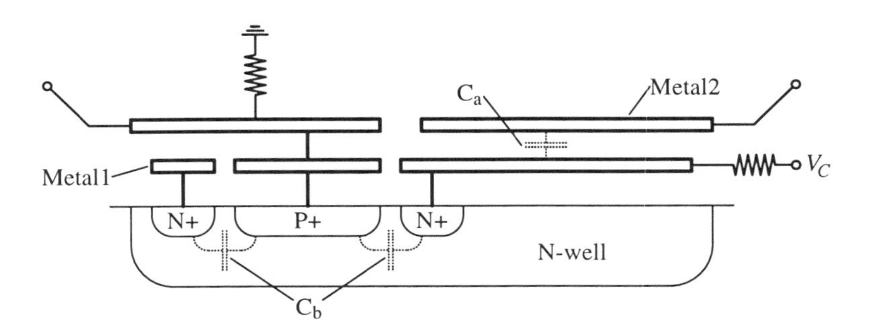

Figure 4.18. Capacitor structure for the enhanced bonding wire LC-tank

a proper series resistance as was deducted from the finite-element simulations. All bonding pads are replaced with the equivalent circuit of figure 4.11. The capacitor parasitics are the series resistance and the parasitic capacitance to the substrate. The series resistance of the junction capacitor is rather large, because of the large N-well sheet resistance. To diminish this effect, the layout consisted of many small capacitors in parallel. However, the junction capacitor series resistance still has the largest contribution to the total resistance. Incorporating an estimation for the transistor parasitic capacitances, these simulations yield an oscillation frequency of approximately 1.8 GHz. The actual oscillation frequency can only be determined when the amplifier transistors are sized and their parasitics are known exactly. The effective resistance of the complete structure with all the parasitics included is derived from calculations and simulations, and equals 3 Ω. The effective capacitance is 3 pF. Therefore, according to equation (3.103), the necessary transconductance of the amplifier will be :

$$G_M = 3\Omega \times (2\pi \cdot 1.8GHz \times 3pF)^2 = 3mS \tag{4.10}$$

For an oscillation amplitude of 1.8 $V_{diff,peak}$, and assuming a value of 3 for the amplifier excess noise factor A, the expected phase noise is :

$$\mathcal{L}\{200kHz\} = \frac{kT \cdot 3\Omega \cdot [1+3] \cdot \left(\frac{1.8GHz}{200kHz}\right)^2}{\frac{(1.8V)^2}{2}} = 2.5 \cdot 10^{-12} = -116dBc/Hz \tag{4.11}$$

Another advantage of this type of LC-tank is the branch having only one inductor (L_1). This makes it possible to develop a simple transconductor, since the tank's impedance at low frequencies is zero. Oscillators employing another kind of enhanced LC-tank, such as a crystal oscillator, need to take special care of the DC-biasing of the amplifier. Usually two large capacitors are placed in parallel with the crystal [Vitto JSSC88]. For the high frequencies used in personal communications, the series resistance associated with these capacitors severely decreases the obtained transconductance. Amplifiers employing only one or two transistors are preferred.

4.4 AMPLIFIER DESIGN

4.4.1 Circuit Schematic

A general rule for high-frequency oscillator schematics is : keep it simple! Every extra transistor will add parasitic capacitances and/or resistances that will eventually limit the circuit performance. Since the only function the active circuit must perform, is to create a negative resistance in parallel with the LC-tank, very few transistors are necessary. This can also be seen in the crystal oscillator schematics in figure 2.18 which are based on only one transistor.

The amplifier used in this design is a common source structure, allowing very high operating frequencies. The circuit diagram is shown in figure 4.19. The PMOS transistors act as current sources, so the actual amplification is done with only two NMOS transistors, $M1$ and $M2$. These two transistors are cross-coupled to provide the required positive feedback. The realized transconductance G_M is half of the transconductance of one NMOS transistor.

This circuit configuration can operate at very high frequencies. Indeed, there is no f_T limitation since the transistors' gate-source capacitance is situated in parallel with the capacitors of the LC-tank. The inductor is used to charge $C_{GS,M1,M2}$, so there is no pole g_m/C_{GS}. The only limitation present is caused by the transistor's drain resistance and capacitance. Especially the drain series resistance can become unexpectedly high in submicron processes, since the lowly-doped-drain (LDD) structures used to prevent

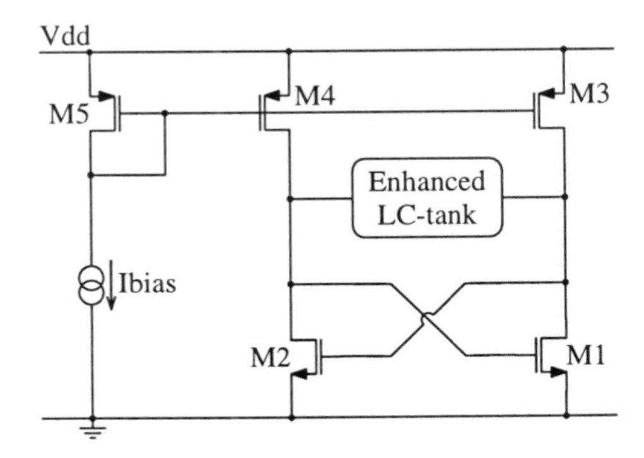

Figure 4.19. CMOS bonding wire VCO circuit schematic

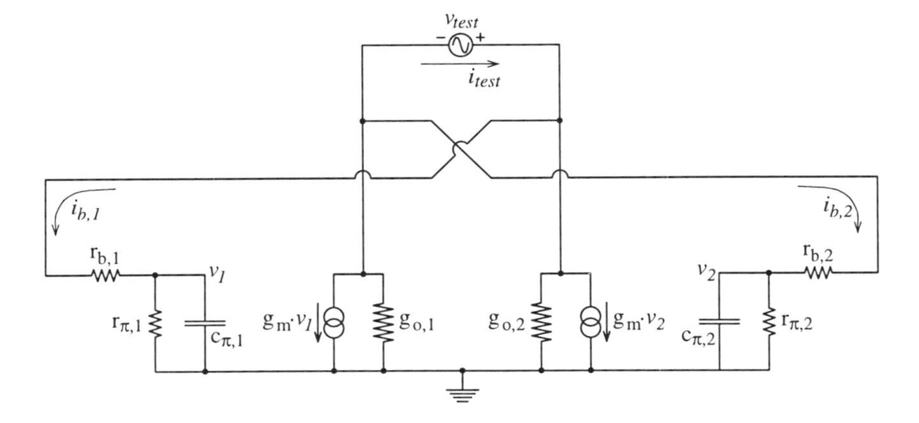

Figure 4.20. Bipolar negative resistance frequency limit

hot-electron currents represent a large resistive path in the drain. From calculations and simulations the frequency limitation is expected to be in the order of 4 to 5 GHz.

4.4.2 Bipolar or CMOS ?

In a bipolar technology, high-frequency VCO design is more difficult. This is caused by the large base resistance r_b that forms a frequency-limiting pole together with the small-signal base-emitter capacitance c_π. We can calculate a theoretical maximum frequency for a common source bipolar amplifier as follows. In figure 4.20 the small-signal equivalent model of two cross-coupled bipolar transistors in positive feedback is depicted. A very simple transistor model is used, consisting only of the base resistance r_b, base-emitter capacitance c_π and resistance r_π, transconductance g_m and output conductance g_o. A voltage source v_{test} is applied and the negative resistance created will be deducted from the resulting current i_{test} through this source.

The resulting admittance is

$$Y(s) = -\frac{1}{2} \cdot \left(\frac{g_m r_\pi - 1 - s r_\pi c_\pi}{r_b + r_\pi + s r_b r_\pi c_\pi} - g_o \right) \tag{4.12}$$

Assuming $g_m r_\pi = \beta \gg 1$, $r_\pi \gg r_b$, and neglecting g_o, this can be simplified to

$$Y(s) \approx -\frac{g_m}{2} \cdot \frac{1 - s c_\pi / g_m}{1 + s r_b c_\pi} \tag{4.13}$$

At low frequencies, the impedance equals -$g_m/2$, as expected. At higher frequencies, the circuit can only function as long as the real part of the realized impedance remains smaller than zero. Substituting $s = j \cdot \omega$, this real part is calculated to be

$$\Re\{Y(\omega)\} = -\frac{g_m}{2} \cdot \frac{1 - \omega^2 r_b / g_m c_\pi^2}{1 + (\omega r_b c_\pi)^2} \tag{4.14}$$

which can only be negative for frequencies smaller than

$$\omega_{max} = \frac{\sqrt{g_m / r_b}}{c_\pi} \tag{4.15}$$

This clearly illustrates the need for a low base resistance in high-speed bipolar transistors, a fact that is often neglected by technology developers when optimizing the f_T of the process. The f_T of a transistor is the frequency at which the current gain drops below 1, and is in first order equal to

$$f_T = \frac{g_m}{2\pi \cdot c_\pi} = \frac{g_m}{2\pi \cdot (C_{jBE} + \tau_F g_m)} \tag{4.16}$$

with C_{jBE} the intrinsic junction base-emitter capacitance and τ_F the forward base transit time [Laker 94]. This forward transit time can be minimized by using a very short base width. This results in a high f_T, but also in large base resistance, which is a disadvantage for high-speed circuit design. The maximum frequency of oscillation can be written as a function of f_T :

$$f_{max} = \frac{1}{2\pi} \omega_{max} = \frac{1}{\sqrt{g_m r_b}} \cdot f_T \tag{4.17}$$

For example, in the bonding wire test VCO of section 4.2.2 the base resistance is as high as 1 $k\Omega$. With a current of 0.5 mA per transistor, a transconductance g_m of 20 mS is obtained. So oscillation is only possible at frequencies smaller than one fifth of f_T. With more complicated transistor models, the actual maximum frequency will be even smaller. Since the gate resistance of CMOS transistors can always be made arbitrarily small by using a large number of fingers in parallel, CMOS oscillators do not suffer from this frequency limit.

4.4.3 Circuit Sizing

The transistors $M1$ and $M2$ are biased such that the gate voltages are approximately half-way the 3-V power supply. They therefore have $V_{GS} - V_T \approx 0.7V$. This leads to a very large linear range for the transconductor, allowing a large oscillation amplitude and thus a low phase noise. The realized negative conductance is chosen to be twice as large as the value required by (4.10), or 6 mS. This requires a current

$$I_{M1,2} = \frac{g_m \cdot (V_{GS} - V_T)}{2} = \frac{12mS \cdot 0.7V}{2} = 4\ mA \tag{4.18}$$

Their length is minimal, i.e. 0.7 μm, and the necessary width is 180 μm. The PMOS current sources $M3$ and $M4$ have a larger length of 1.2 μm, to increase their output resistance, and a width of 720 μm. The circuit is biased through $M5$ and an external current source.

The drain-source current noise of a single transistor with transconductance g_m equals $di_{DS}^2 = 4kT \cdot 2/3 \cdot g_m \cdot n \cdot df$, with n a correction factor incorporating the modulation of the bulk transconductance by channel thermal noise [Enz AICSP95]. n typically has a value equal to 1.5. In the VCO amplifier, both the NMOS amplifying transistors and the PMOS current sources contribute to the output current noise. Since the PMOS transconductance approximately equals the NMOS transconductance, the output current noise will be multiplied by a factor of 2. Finally, the total transconductance was two times larger than strictly necessary, which can be incorporated by making α equal to 2. This leads to the following value of A :

$$A = \alpha \cdot F_{G_M} = 2 \times (2/3 \cdot 1.5 \cdot 2) = 4 \tag{4.19}$$

With this correct value for the factor A, we can adjust equation (4.11) to the final value of $\mathcal{L}\{200\, kHz\} = -115\, dBc/Hz$.

4.5 MEASUREMENT RESULTS

A microphotograph of the IC is shown in figure 4.21. The on-chip bonding wires can clearly be seen. The VCO signal feeds a capacitive voltage divider, the output of which is approximately 40 dB smaller than the VCO signal. This reduced signal can be connected to the 50-Ω spectrum analyzer without deteriorating the quality of the LC-tank. The VCO output spectrum is shown in figure 4.22(a). The center frequency is 1.76 GHz, the resolution bandwidth is 10 kHz and the video bandwidth is 30 Hz. The measured center frequency is only 2% off from the calculated one. The chip-to-chip variation in center frequency has a σ-value less than 1%.

A logarithmic plot of the oscillator phase noise as a function of frequency offset from the carrier is shown in figure 4.22(b). The measured single sided spectral phase noise density at 200 kHz from the carrier is -115 dBc/Hz. This result is exactly equal to the expected performance. At 10 kHz from the carrier, the measured phase noise is -85 dBc/Hz. This does not agree anymore with the quadratical decrease in phase noise with the frequency offset, but is due to a small influence of the modulation by $1/f$-noise at frequencies very close to the carrier.

This measurement was done with the most simple and cheapest method, i.e. a spectrum analyzer. But due to the limited dynamic range of the system, phase noise below e.g. -120 dBc/Hz cannot be measured. This can be seen in figure 4.22, where the straight curve already flattens off for larger offset frequencies. It is thus not possible to measure the phase noise at a few MHz offset. Another disadvantage is the fact that this setup cannot make a distinction between phase noise and amplitude noise, as

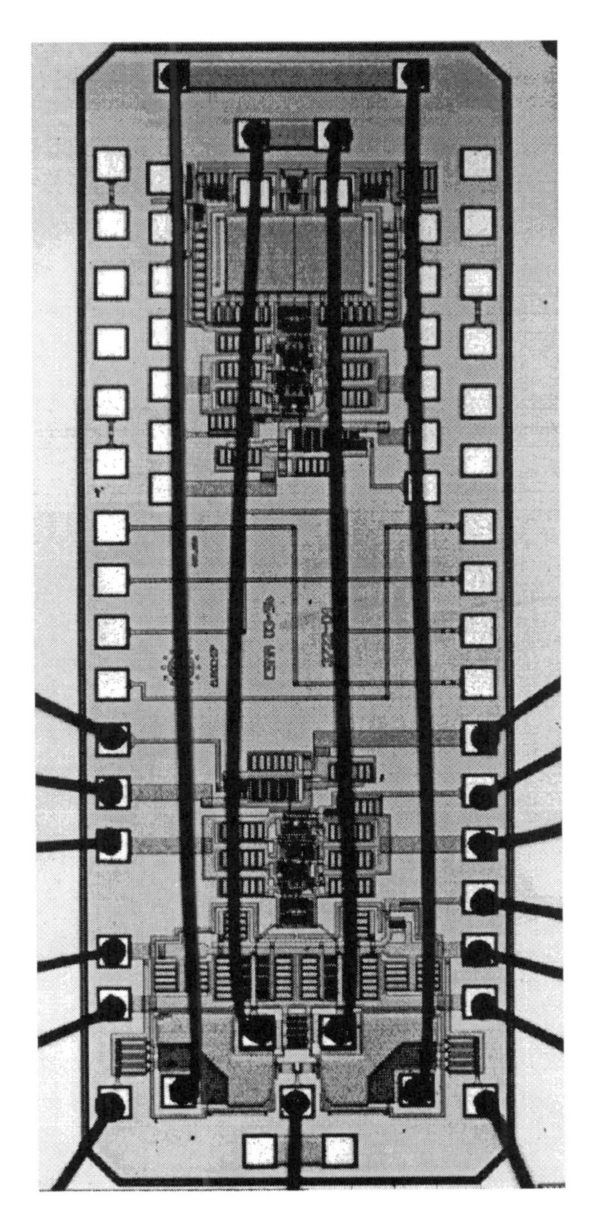

Figure 4.21. IC microphotograph of CMOS bonding wire VCO

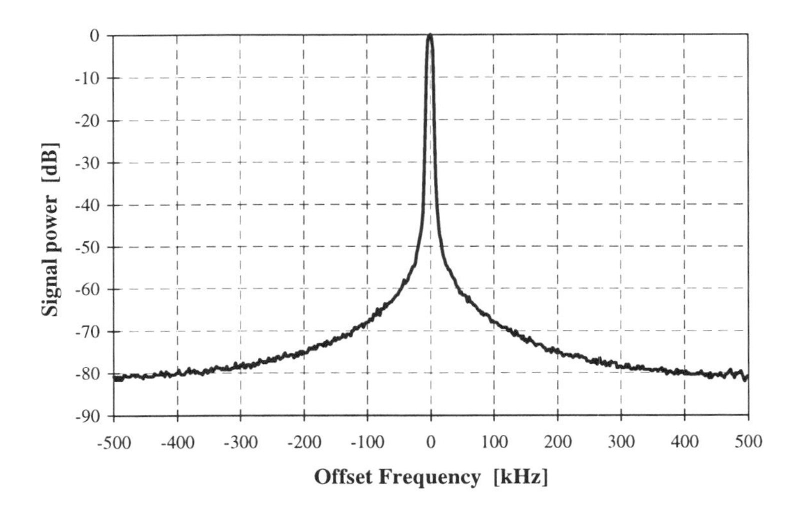

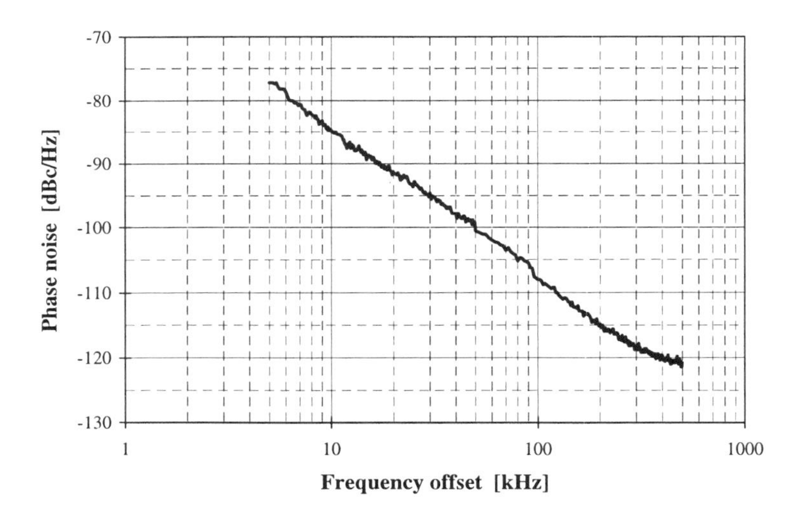

Figure 4.22. Measured phase noise of the CMOS bonding wire VCO : (a) Output spectrum for a carrier frequency of $1.76\ GHz$ and resolution bandwidth of $10\ kHz$; (b) Logarithmic plot

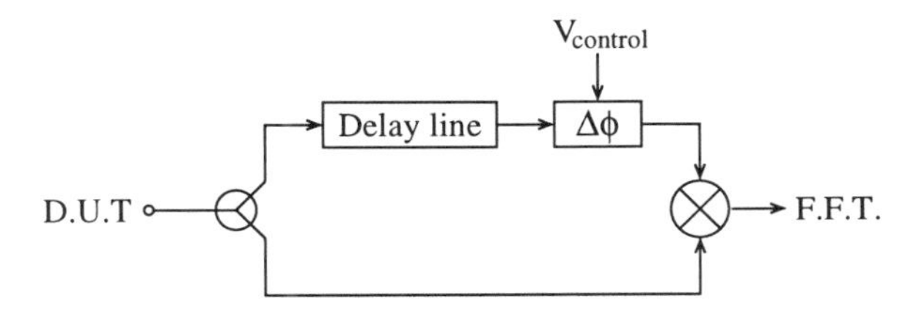

Figure 4.23. Phase noise measurement with delay line system

explained in chapter 3. The measured noise power density is the sum of AM and PM noise, so the pure phase noise of the oscillator will always be lower than measured by the spectrum analyzer.

A dedicated phase noise measurement setup is shown in figure 4.23. The DUT (Device Under Test) signal is split in two in a power splitter. One path goes straight to the mixer, but the other signal first passes through a delay line and a phase shifter. The implemented delay is typically 100 $nsec$, and loss in the delay line must be compensated by the proper amplification. The phase shifter actually also implements a small delay, but this is controlled by the voltage $V_{control}$. The phase shift is adjusted such that the two input signals to the mixer are in quadrature, and the mixer gain is at its maximum. The two signals are downconverted to baseband in the mixer, and the spectrum of this baseband signal is determined by an FFT. The delay line decorrelates the phase noise in the two signal paths, so the mixer output is 3 dB higher than the phase noise of the DUT. Since the mixer is insensitive the amplitude noise, only the PM noise is downconverted and the FFT spectrum is pure phase noise. Dedicated but expensive measurement systems exist that implement these functions and are controlled by e.g. a personal computer to adjust the phase shift, calibrate the system and calculate the phase noise. A measurement of the bonding wire VCO with such a system is shown in figure 4.24.

Two important conclusions can be drawn from these measurements. First, with respect to the spectrum analyzer measurements there is an improvement by approximately 4 dB, since the measured value at 200 kHz offset is now -119 dBc/Hz. This difference is due to the distinction between PM noise and AM noise that is made by this setup. Theoretically, there can only be an improvement by a factor of two (= 3 dB), but the larger measured improvement can be attributed to the overall measurement accuracy of 2 dB. Secondly, the theoretical 20-dB/dec slope continues up to a least a few MHz. The measured phase noise at 3 MHz offset equals -146 dBc/Hz.

The oscillator operates on a single 3-V power supply and consumes 8 mA. The tuning characteristic is shown in figure 4.25. The tuning range is 4.5% for tuning

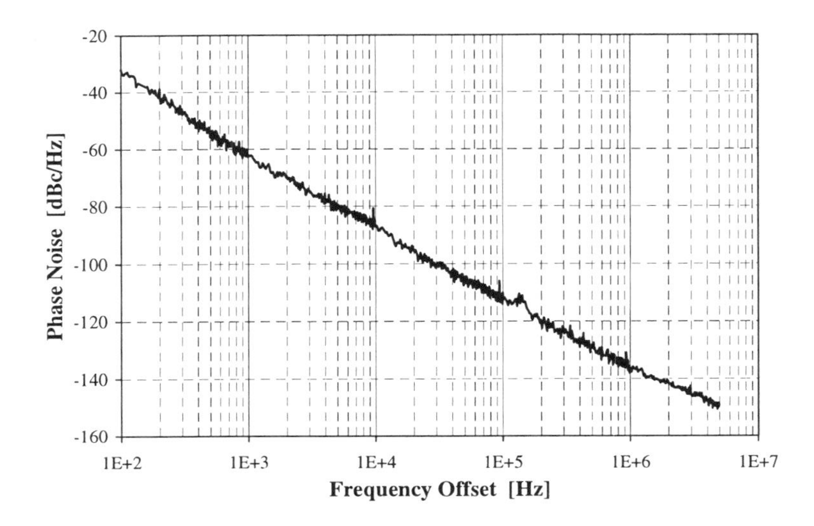

Figure 4.24. Measured CMOS bonding wire VCO phase noise with the delay line method

voltages up to 3 V and almost 7% for tuning voltages up to 10 V. If processing variations or other specifications require a larger tuning range, this can be done by digitally adding/removing capacitance in discrete steps [Dunca CICC95, Parke CICC97].

4.6 CONCLUSIONS

In this chapter we have investigated the use of bonding wire inductors in a fully integrated VCO. The low series resistance of gold bonding wires promises the possibility of high quality factors and hence low phase noise. We have studied the peculiarities of bonding wires in detail to see if that promise comes true. The first-order formulas for inductance calculations given in [Green TPHP74] are extremely accurate at low frequencies, as was demonstrated by comparison with finite-element simulations. They show that several nH of inductance value can be achieved. But two aspects will limit this ideal situation.

First, the bonding process is not 100% reproducible and the shape of the wire will vary from chip to chip. This causes an uncertainty in the final inductance value obtained, which must be incorporated in the total tuning range of the VCO. The parameters that characterize the bond are :

- the height above the substrate
- the vertical bend in the bonding wire, which causes a change in the self-inductance

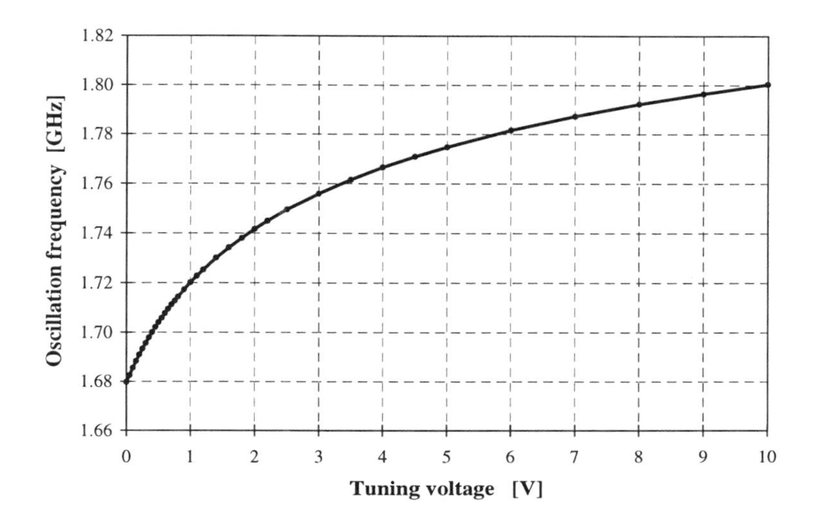

Figure 4.25. Measured CMOS bonding wire VCO tuning characteristic

- the horizontal bend in the bonding wire, which causes a change in the mutual inductance between two wires
- the wire diameter

A qualitative and quantitative study of the influences of the variations in all these parameters has been done, based on which we can state that we can safely expect the actual inductance value to be within 6 % of the expected value.

Secondly, there are the parasitic elements of the inductor, which are mostly situated in the wire resistance and the bonding pads. The bonding pads have a parasitic capacitance to the substrate, and an associated series resistance. The actual values of these elements depend on the bonding pad structure, the wafer type, and the layout of e.g. substrate contacts around the pad. The wire series resistance suffers to a great extent of current redistribution due to the skin effect, and hence the resistance increases significantly at high frequencies. Another element that must be taken into consideration, are the losses in the substrate if this is heavily doped.

We have applied all the gathered knowledge about these bonding wires in the design of a fully integrated high-frequency VCO. An enhanced LC-tank was formed with two differential bonding wire inductors and two tunable junction capacitors. A simple CMOS negative resistance was added to create the instability required for oscillation. CMOS is the best suited technology for this application, since bipolar transistors suffer from a frequency limitation due to the pole formed by the base resistance and the small-signal base-emitter capacitance.

The resulting VCO has a 1.8-GHz center frequency, consumes 24 mW, and has a phase noise of -115 dBc/Hz at an offset of 200 kHz. As can be seen from the data in table 3.2, this is the lowest phase noise obtained by any integrated oscillator published in open literature. Although other designs use exotic processing techniques such as substrate etching or thick metalization, this bonding wire VCO is 5 dB better. This improvement is partly due to the high-quality bonding wires, and to the power-noise trade-off in the enhanced LC-tank.

Although the statement that bonding wires are indeed integrated inductors might be somewhat controversial, is it clear that they offer better performances than planar inductors. Although there are some improvements possible to the spiral-inductor LC-oscillators currently found in open literature, as is demonstrated in the next chapter, there are applications where the phase noise specs are too tough. In this case the designer should consider the use of bonding wire inductors, and make the cost trade-off between the extra design effort and large silicon area over external components.

5 PLANAR-INDUCTOR VCOS

5.1 INTRODUCTION

After the discussion of bonding wire inductors in the previous chapter, we will now investigate the more "normal" type of integrated inductor, i.e. the spiral or planar inductor. This is indeed the dream of every RF VCO designer : be able to use a simple spiral metal track laid out in the standard metal routing levels of the IC as an inductor. If the required specifications can be achieved with this technique, the cost will certainly be lower than a design using bonding wire inductors. The die area can be reduced by an order of magnitude, and the variations in inductance value due to non-uniform bonding are of course no longer present. Once the processing masks are made, the inductor geometry will always be the same, since the processing accuracy ($< 0.1\mu m$) is certainly negligible with respect to the size of the coil ($\approx 100\mu m$). So the inductance value variations do no longer have to be compensated for by the VCO tuning range.

Although the planar-inductor solution is certainly more elegant than the bonding-wire design, it has some major drawbacks. A lot of parasitics limit the quality of the inductors, which has led many designers to the use of special processing techniques to enhance their performance. To decrease the series resistance, multi-level and extra thick metalization are used. And to eliminate the self-resonance with the parasitic capacitance to the substrate, this substrate is etched away underneath the inductor. But now the low-cost advantages of integrated planar inductors with respect to external components have vanished due to this extra processing. If one wants to integrate the RF front-end of a mobile transceiver with the baseband signal processing, the technology used has to be the cheapest one possible, i.e. plain standard digital CMOS. No tuning, trimming, extra processing or etching steps are allowed. This is the starting point in the investigations in this chapter : how far can we go in the phase noise of a planar-inductor LC oscillator in a standard CMOS process, with no extra cost allowed ?

The first section of this chapter will study the planar inductors on a silicon substrate in great detail. The first-order equations for the inductance calculation are given, but they are not sufficient to incorporate all high-frequency magnetic field effects. Therefore finite-element simulations are necessary, for which a very efficient simulation strategy is developed that reduces the complexity of the problem to two dimensions by employing the symmetry around the vertical axis. A thorough analysis of these simulation results leads to some general guidelines for planar inductor sizing, the most important of which is the fact that the best performance is obtained when a hole is left in the middle of the coil where no conductor tracks are placed, i.e. the coil must be "hollow".

The next two sections discuss two designs that use planar inductors that were optimized for the application in a low-phase-noise oscillator. Section 5.3 presents the first design, which is done in a 0.7-μm CMOS technology, which uses epi-wafers and has only two metal layers. The finite-element simulations allow to size the inductors used to an optimum, thereby surpassing by many deciBells the phase noise of other comparable designs published in open literature. The second design is discussed in section 5.4. It consists of a double design, one at 900 MHz and one at 1.8 GHz. The 0.4-μm CMOS technology of this design uses lowly-doped wafers and since this limits the losses in the silicon substrate, another improvement in phase noise is possible. No extra processing or etching steps are used, and yet these VCOs achieve the required phase noise specification for the GSM and DCS-1800 systems.

5.2 PLANAR INDUCTORS

Planar inductors consist of a rectangular spiral metal track on the silicon substrate, using one or more of the standard metal interconnection levels available, as shown in figure 5.1(a). The basics for the inductance calculation of these planar inductors were developed by Greenhouse in 1974 [Green TPHP74]. If the technology allows 45-degree routing, an octagonal shape can be used (figure 5.1(b)), but for this structure the first-order formulas are no longer applicable. Before presenting this first-order

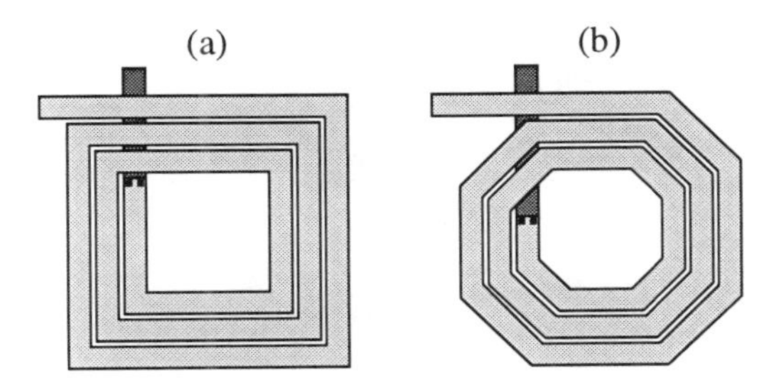

Figure 5.1. Silicon spiral inductor layout : (a) square; (b) octagonal

inductance model, we will shortly discuss the most obvious parasitics of the planar inductor.

The quality factor of the inductor will be limited by the series resistance of the metal tracks. Typical values in a simple process are 15 Ω for a 10-nH inductor [Nguye JSSC90]. GaAs circuits, which incorporate a lot of planar inductors, have the advantage of gold interconnects to achieve low series resistance. In silicon, most designers try to use very wide metal turns to compensate for the higher sheet resistance of the metalization used [Based ESSC94, Soyue JSSC96a]. However, we will show that due to high-frequency magnetic field effects (such as the well-known skin effect) this is not necessarily the best solution.

One of the most recognized parasitics is the capacitance to the substrate. Together with the wanted inductance, this gives an LC resonance frequency above which the coil can no longer be used as an inductor. A typical value for the self-resonance frequency is 2.5 GHz for a 10-nH inductor [Nguye JSSC90]. This puts a limit on the maximal inductance value achievable at a certain frequency, as larger inductors require a larger area and thus also a larger capacitance and a smaller self-resonance frequency. Special processing technologies exist that create an air-gap underneath the inductor, or have very thick oxide under their top metal routing level [Soyue JSSC96b]. The resulting smaller parasitic capacitance allows higher operating frequencies. As already mentioned in the introduction, more recently another solution for the self-resonance problem has been developed. By using selective etching techniques, the silicon substrate can be removed in a post-processing step from underneath the inductor, either by etching from the top of the wafer [Chang EDL93, Rofou ISSCC96], or from the back of the wafer [Based ESSC94]. An example of such an etched inductor is shown in figure 5.2.

During layout, masks are drawn such that holes in the field oxide next to the inductor remain after processing. This done by e.g. drawing a contact and a via hole over that area (this etches the oxide away before metalization), and indicating that no passivation must be placed, as is also done for bonding pads. The wafers are then

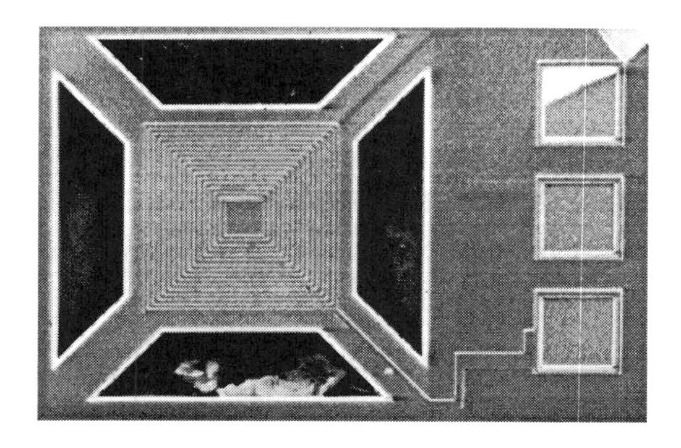

Figure 5.2. Etched silicon spiral inductors : (a) Top view; (b) Cross section

submitted for normal processing, and afterwards the etching step is done on naked dies. The isotropic etchant removes all the silicon material underneath the inductor, but does not touch the oxide or the metalization. That way only the inductor remains, supported by a field oxide bridge across the silicon gap. This structure should be strong enough to withstand sudden shocks during handling of the ICs. Care must be taken during the design of the oxide bridge structure, because stress in the remaining oxide can cause the structure to crack [Olgaa NORC93]. Bringing this technique into a mass-production cleanroom is thus not straightforward. Also, not every process can be adopted for these etched inductors. Some new processing steps or new materials used in modern submicron processes might even be incompatible with the etching technique. An example of this is the currently often used tungsten fill of contacts and vias. With this technique the contact hole that was drawn during layout does not remain empty but is filled up with another material such as tungsten. So post-processing etching is impossible.

The substrate etch removes the capacitance to the substrate, and shifts e.g. the self-resonance frequency of a 100-nH inductor from 0.8 GHz to 3 GHz [Chang EDL93]. But more important, this technique also eliminates the resistive losses in the substrate. These losses are indeed the most important limitation in using planar inductors in a standard CMOS process. As in section 4.2.4.1, a distinction must be made between bulk- and epi-wafers. The bulk wafers have a lowly-doped substrate, resulting in a substrate resistivity in the order of 10 Ωcm. The epi-wafers have a substrate resistivity in the order of 0.01 Ωcm, which is why now currents in the substrate, that are generated by the magnetic field of the inductor, are free to flow. This severely increases the losses and reduces the inductance value.

We will now come back to our original goal : what specs can we achieve in a standard CMOS digital process ? Only two metal layers are available, and the losses in the substrate have to be taken into account as no post-processing is allowed. We will now start studying these structures and analyze all parasitic effects.

5.2.1 First-Order Inductance Calculation

A general reference for the calculations of planar inductors is [Green TPHP74]. This article presents a method to accurately calculate the inductance value that is based on summing/subtracting the appropriate self- and mutual inductance values of the different segments of the coil. Since its publication, it has been cited many times as a reference by anyone using planar inductors in his design. The derivations start from basic formulas for the exact self inductance of a straight conductor :

$$L = \frac{\ell}{5} \cdot \left[\ln\left(\frac{2\ell}{GMD}\right) - 1.25 + \frac{AMD}{\ell} + \frac{\mu_r}{4} T \right] \tag{5.1}$$

where L is the self inductance in nH, ℓ is the conductor length in mm, GMD and AMD are the geometric and arithmetic mean distances, respectively, of the conductor cross section in mm, μ_r is the relative permeability of the conductor, and T is a frequency-correcting parameter.

The geometric mean distance (GMD) between two conductors is the distance between two infinitely thin imaginary filaments whose mutual inductance is equal to the mutual inductance between the two original conductors. By definition, the self inductance of a conductor is the sum of all the mutual inductances of all the pairs of filaments of which it is composed. Calculation of the GMD is lengthy, but for a rectangular cross section of width w and height h it equals $0.22313 \cdot (w + h)$ [Green TPHP74]. The arithmetic mean distance (AMD) is the average of all the distances between the points of one conductor and the points of another. For a single conductor, it is the average of all possible distances within the cross section. In the case of a rectangular cross section with small height h, as in planar inductors, the AMD can be approximated by the AMD of a straight-line cross section, i.e. one third of the sum of width and height. Assuming $\mu_r = 1$ and $T = 1$, (5.1) reduces to

$$L = \frac{\ell}{5} \cdot \left[\ln\left(\frac{2\ell}{w+h}\right) + 0.50049 + \frac{w+h}{3\ell} \right] \tag{5.2}$$

The mutual inductance between two parallel conductors is a function of the length of the conductors and of the geometric mean distance between them :

$$M = \frac{\ell}{5} \cdot \left[\ln\left(\frac{\ell}{GMD} + \sqrt{1 + \left(\frac{\ell}{GMD}\right)^2}\right) - \sqrt{1 + \left(\frac{GMD}{\ell}\right)^2} + \frac{d}{\ell} \right] \tag{5.3}$$

where GMD is now the geometric mean distance (GMD) between the two conductors and is approximately equal to the distance d between the centers of the tracks. The exact value of GMD must be calculated from

$$\ln GMD = \ln d - \left[\frac{1}{12}\left(\frac{d}{w}\right)^2 + \frac{1}{60}\left(\frac{d}{w}\right)^4 + \frac{1}{168}\left(\frac{d}{w}\right)^6 + \frac{1}{360}\left(\frac{d}{w}\right)^8 + \cdots \right] \tag{5.4}$$

Calculation of the total inductance value of a spiral coil with many turns then involves the calculation of every self inductance and of every mutual inductance between every possible couple of segments. The mutual inductance can be positive (M^+) if the current in the segments is flowing in the same direction, or negative (M^-) if current flow is in the opposite direction. Orthogonal segments do not have any mutual inductance contribution. The total inductance value is than given by

$$L = \sum_i L_i + \sum_{i,j} M_{i,j}^+ - \sum_{i,j} M_{i,j}^- \tag{5.5}$$

Because of the large number of calculations involved, a custom computer program can be used [Green TPHP74]. The reported accuracy of the calculated inductance value is satisfying. But since the mutual-inductance formulas are not valid for segments with a 45-degree angle, this technique is not applicable for the octagonal inductor of figure 5.1(b).

5.2.2 Finite-Element Simulations

The biggest problem of this first-order inductor model is the fact that it does not take into account all parasitics of the inductor when this one is used at high frequencies on a conductive silicon substrate. From the discussion in the beginning of this section, we can conclude that spiral inductors suffer from three parasitic effects.

- First, parasitic capacitance to the substrate causes the inductor to self-resonate at a certain frequency. This capacitance value is easily calculated and poses no problems for simulation.

- A second and more important parasitic is the inductor series resistance, which will differ at high frequencies from the calculated one due to the skin effect and other magnetic field effects. Basically these effects will cause the current flow through the conductor track to be non-uniform and hence change the resistance value. At first, the basic formulas cannot handle this, but an extension has been proposed in [Nikne CICC97], which splits up each conductor track into many parallel segments. The interactions between the self- and mutual inductances of these elements allow to take the current crowding effects into account.

- Thirdly, the losses in the heavily doped substrate cause a large degradation in the overall quality factor and reduce the inductance value. This cannot be included by methods described above, because it is a distributed effect that cannot be captured by the interactions between discrete conductors.

In order to better understand all these effects, an efficient finite-element simulation strategy is developed. Simulation is necessary because the problem is much too complex to be solved analytically. The only way to fully incorporate all the effects is a full 3-D finite-element simulation. That is however very time-consuming. We want to gain some insight in the severity of the several parasitic effects as a function of coil geometry by simulating many different coils. This should lead to some general design guidelines for planar inductors. Therefore, simulation must be done as fast as possible. Planar 2-D or so-called 2.5-D simulators work very fast, and can operate on complex coil geometries. But these are not sufficient for the problem, because they do not account completely for the substrate effects.

We have chosen to simulate circular inductors instead of square coils. Circular inductors offer the advantage of being symmetrical around the vertical axis, and the problem therefore only has a 2-D complexity. 2-D Simulation can be done very fast. The simulated structure is shown in figure 5.3. The coil has n metal turns, an outer radius r, conductor width w and spacing sp between the conductors. The substrate consists of a heavily doped bulk material, a lowly doped epi layer, and an n-well (or p-well) layer. On top of this is the field oxide. A current of a certain frequency is forced through the coil, and the resulting magnetic field and all resistive losses are calculated. This simulation was done with the same program as the bonding wire simulations in chapter 4, i.e. MagNet [Freem 93] and takes approximately 10 to 15 minutes per inductor. This gives us a very fast way to calculate the inductance and

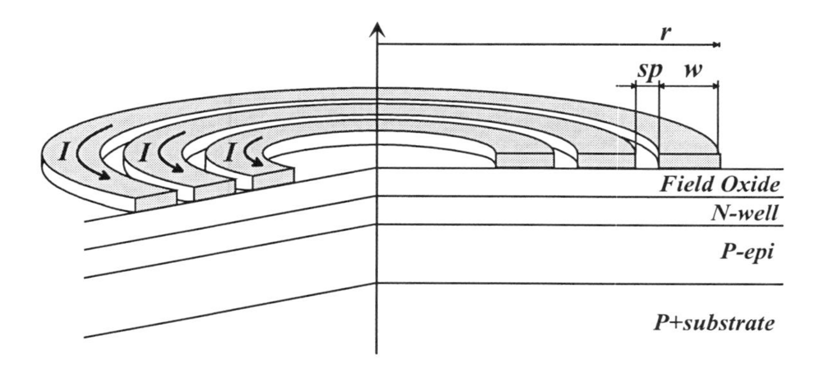

Figure 5.3. Cross-section of the circular inductor model for finite-element simulations

the equivalent series resistance of several coil geometries. We can make a distinction between losses in the metal conductors and losses in the silicon. That way we can gain insight in the several parasitic effects in the metal conductors and in the substrate. The conclusions drawn from these simulations are now discussed.

5.2.2.1 Metal Losses. At low frequencies, the series resistance of the metal conductor tracks can easily be calculated as the product of the sheet resistance and the number of squares of the metal track. At high frequencies, however, the skin effect and other magnetic field effects will cause a nonuniform current distribution in the inductor. This will have a (sometimes serious) influence on the losses in the metal conductor at high frequencies.

The best known of these effects is the skin effect. It can be analyzed analytically for a straight metal conductor with circular cross section, as was done in section 4.2.4.2. In a planar inductor, this effect can no longer be calculated analytically, but it is clearly seen in the finite-element simulations. As an example, two inductor geometries are simulated, one with parameters $n = 2$, $r = 116\mu m$, $w = 15\mu m$, $sp = 2\mu m$ and one with parameters $n = 2$, $r = 151\mu m$, $w = 30\mu m$, $sp = 2\mu m$. So the second inductor has metal turns which are twice as wide as the first one. The radius has been adjusted in order to achieve approximately the same inductance value. The wide metal turns should allow a low series resistance and hence low phase noise in the oscillator. Due to the larger radius r, the resistor ratio is not as large as two, but equals 1.67. Since the phase noise of an LC-oscillator is proportional to the equivalent series resistance of the LC-tank (see chapter 3), a $10 \cdot \log(1.67) = 2.2\ dB$ better phase noise performance should be possible with the second coil. Figure 5.4 shows the variation of the simulated metal series resistance as a function of frequency. The ratio $Rs/R@dc$ is drawn, i.e. the ratio between the effective series resistance at a

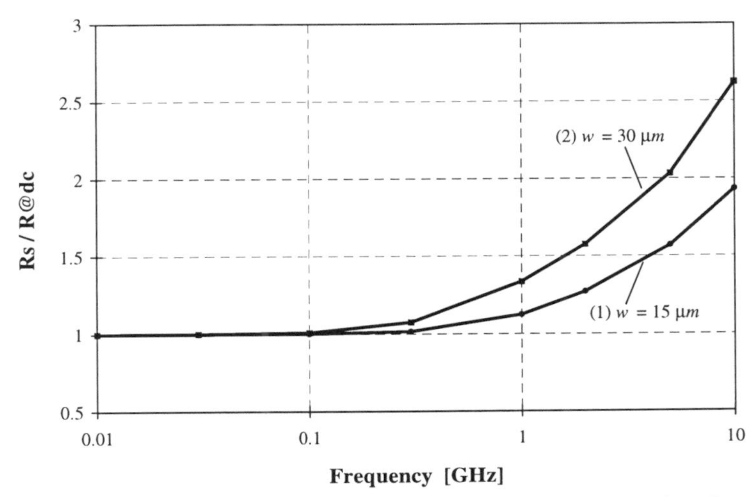

Figure 5.4. Simulation results for the influence of the skin effect on the planar inductor series resistance; both inductors have two turns and equal inductance value, but different conductor width

certain frequency and the resistance at DC. At 2 GHz the series resistance of the second inductor (2) is already 60% higher than the value at DC, while the first one (1) only suffers a 30% increase. Therefore, the difference in resistance between the two inductors is now only a factor 1.35, or 1.3 dB. The inductance value remains roughly unchanged for both. At even higher frequencies, the resistance increase is enormous.

This proves that inductors using very wide metal turns are not the way to go in designing low-phase-noise oscillators. Due to the larger required coil radius when using wide turns to maintain the inductance value, the phase noise does not decrease linearly with conductor width. This is even enhanced by the skin effect which seriously deteriorates the series resistance of wide conductors at high frequencies.

If we look at the individual resistance of each metal turn, we notice another, even more important, effect. Normally, we should expect a high resistance at the outer turns, because they are the longest, which then gradually decreases for the inner turns. However, an unexpected dependency results from the finite-element simulations. An inductor with parameters $n = 9, r = 100\mu m, w = 9\mu m$ and $sp = 1\mu m$ was simulated at several frequencies. Table 5.1 gives, as a function of frequency, the inductance value L, the total metal series resistance R_{Al} and the individual resistance $R1..R9$ of each turn. $R1$ is the resistance of the outer turn, $R9$ of the innermost one. Figure 5.5 gives a 3-D graphical representation of the data in table 5.1. This data shows that at high frequencies, the largest contribution to the series resistance does not come from the longer outer turns, but from the inner turns ! $R1$ increases from 1.03 Ω at low

Freq.	*L*	R_{Al}	*Resistance per metal track [Ω]*								
[GHz]	[nH]	[Ω]	*R1*	*R2*	*R3*	*R4*	*R5*	*R6*	*R7*	*R8*	*R9*
0.01	6.84	5.39	1.03	0.92	0.82	0.71	0.60	0.49	0.38	0.27	0.16
0.03	6.84	5.39	1.03	0.92	0.82	0.71	0.60	0.49	0.38	0.27	0.16
0.10	6.84	5.40	1.03	0.92	0.82	0.71	0.60	0.49	0.38	0.28	0.17
0.30	6.84	5.50	1.04	0.92	0.82	0.72	0.61	0.51	0.40	0.30	0.19
1.0	6.80	6.61	1.10	0.93	0.85	0.79	0.73	0.67	0.61	0.52	0.42
2.0	6.68	9.44	1.21	0.94	0.93	0.99	1.06	1.11	1.13	1.12	0.95
5.0	6.39	18.4	1.47	1.03	1.23	1.69	2.16	2.59	2.82	2.99	2.40
10.0	6.18	30.2	1.75	1.20	1.68	2.71	3.68	4.66	5.01	5.43	4.09

Table 5.1. Simulation results for the individual series resistance per metal track; inductor parameters are $n = 9$, $r = 100\mu m$, $w = 9\mu m$ and $sp = 1\mu m$

frequencies to 1.21 Ω at 2 GHz, or by 18%. The increase in $R9$ is from 0.16 Ω to 0.95 Ω, or 480% ! This enormous difference cannot be explained by the skin effect in a single metal track alone, since both tracks are of equal width and they should suffer to the same amount from the nonuniform current distribution.

The cause for this phenomenon can be found in the generation of eddy currents in the inner conductors, as shown in figure 5.6. A part of the right half of the circular inductor is shown schematically, from the outer turn no. 1 to the inner turn no. 9. The inductor carries a current $I_{coil}(t)$, which flows at the depicted instance of time in the counter-clockwise direction. This current of course has an associated magnetic field $B_{coil}(t)$, which has a maximum intensity in the center of the coil. The magnetic field is oriented perpendicular to the page, in the direction coming out of the page (indicated by the symbol ⊙).

When the spiral inductor is filled with turns up to the center of the coil, a large part of the magnetic field does not pass through the center of the coil but goes through those inner turns. Due to the time-varying nature of the coil current, the generated magnetic field $B_{coil}(t)$ also varies with time. According to the law of Faraday-Lenz, an electrical field is magnetically induced in the inner turns that will generate circular eddy currents $I_{eddy}(t)$ as indicated in figure 5.6. The direction of these eddy currents is such that they oppose the original change in magnetic field. So the magnetic field $B_{eddy}(t)$ resulting from the eddy currents has a direction flowing into the page (indicated by the symbol ⊗). The magnitude of the induced electrical field is proportional to the

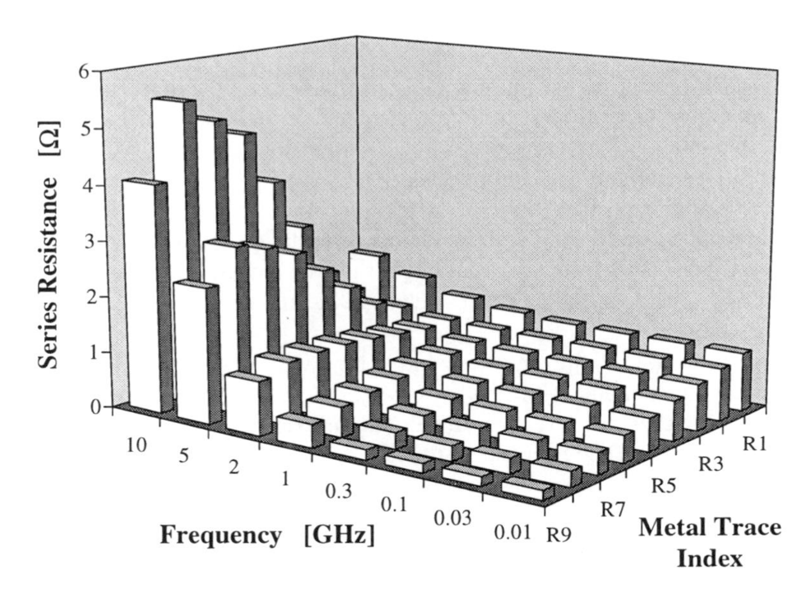

Figure 5.5. 3-D representation of the simulation results for the individual series resistance per metal track

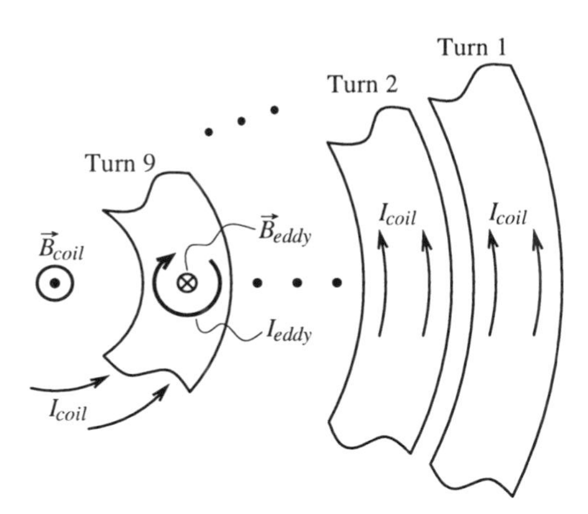

Figure 5.6. Generation of eddy currents in planar inductors

derivative of $B(t)$ to time, so the effect is only noticed at high frequencies. As the total magnetic field ($= B_{coil} + B_{eddy}$) will be smaller, the inductance value will decrease at high frequencies, as noticed in table 5.1.

These eddy currents again cause a nonuniform current flow in the inner coil turns. On the inner side of the inner turn, coil current I_{coil} and eddy current I_{eddy} flow in the same direction, so the current density is larger than average. At the outer side, both currents cancel and the current density is smaller than average. So the current in the inner turns is pushed to the inside of the conductor. This can clearly be seen when analyzing the results of the finite-element simulation. In extreme conditions, the magnitude of the eddy currents is even larger than the coil current, making the current density on the outside of the inner turn negative, i.e. current is flowing in the "wrong" direction.

One might be able to prevent these eddy currents to flow by making longitudinal stripes in the inner conductors, or perhaps by making the inner turns less wide then the outer ones. However, the effect of such countermeasures is questionable, since they will result in a higher DC resistance of the inner turns. These turns already have a low contribution to the inductance, because of the small area they enclose, so even without the eddy currents they cause a (slight) deterioration of the overall quality factor. So it is best if the inner turns are completely omitted, i.e. one should leave a hole in the middle of the spiral coil.

To conclude the discussion of the losses in the metal conductors of a planar inductor, we can safely say that the interactions of skin effect and eddy currents seem far too complex to be analyzed analytically, so the only possible solution to predict the high-frequency metal series resistance is finite-element simulation. As a general rule it can be stated that the conductor width should be limited because of the skin effect, but most important, a "hollow" coil should be used. The inner turns already have a low contribution to the inductance, because of the small area they enclose, and they suffer from an incredible increase in series resistance due to eddy currents at high frequencies. In order to prevent deterioration of the overall quality factor of the inductor, they must simply be left out of the coil.

5.2.2.2 Substrate Losses. As stated earlier, a major drawback of most submicron CMOS technologies is the use of epi-wafers that employ a heavily doped substrate. In these substrates, currents induced by the magnetic field of the inductor are free to flow, which is the cause for extra resistive losses and a decrease in inductance value.

This is shown in figure 5.7. This figure schematically shows a vertical cross-section of the inductor, including the underlying substrate. At the projected instance of time, the inductor current $I(t)$ flows into the page on the right (symbol $\otimes$) and out of the page on the left (symbol $\odot$). As for the eddy currents in the inner conductors, here the law of Faraday-Lenz implies that an electrical field is magnetically induced in an imaginary coil in the substrate underneath the inductor. Therefore, a current $I_{subs}(t)$

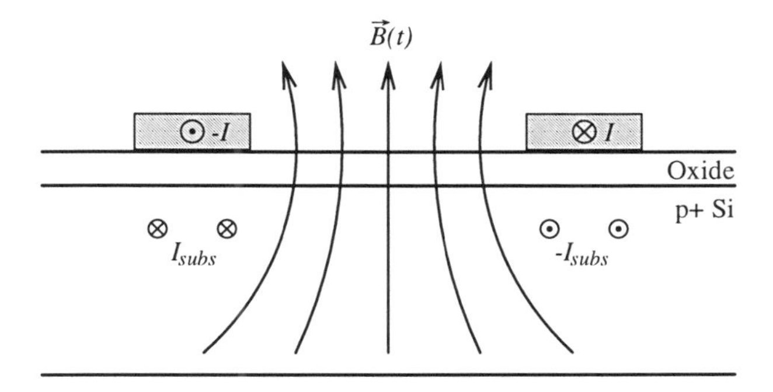

Figure 5.7. Generation of substrate currents underneath planar inductors

will flow in the substrate. The direction of this induced current is such that it opposes the original change in magnetic field. So it flows in a direction opposite to the current in the inductor, as indicated in the figure.

In a substrate with a high resistivity, the induced electrical field only causes a small amount of current I_{subs} to flow, and the effect of the substrate currents can be neglected. This is the case in all GaAs circuits, and in silicon processes that use bulk wafers [Nguye JSSC90]. The quality factor of the inductor is than completely determined by the losses in the metal conductors. Finite-element simulations of coils in such a process indicate that values larger than 10 can be achieved.

In our case, the losses in the heavily doped substrate prevent the realization of such high quality factors. This is demonstrated in figure 5.8. A coil with four turns, having a width $w = 9\mu m$ and spacing $sp = 1\mu m$ on top of a substrate with a resistivity of $0.01\Omega cm$, is simulated at a frequency of 2 GHz for different radii r. Figure 5.8 shows the metal series resistance R_{Al} and the silicon series resistance R_{Si}. This is an equivalent series resistance that models the losses of the induced substrate currents. As the length of the metal track becomes longer, R_{Al} increases gradually with increasing r. The silicon losses show a completely different curve. For small coils, the metal losses dominate, so the quality factor will be determined by R_{Al}. For the largest coil, the metal resistance is 12.1 Ω, whereas the silicon resistance is increased to a value as large as 20.7 Ω. Without the silicon losses, the quality factor would be $2\pi \cdot 2GHz \cdot 13.7nH/12.1\Omega = 14$. This is decreased to only 5 due to the substrate currents. At lower frequencies, the effect is less severe since the changes in the magnetic field are slower.

The inductance value L also suffers from the substrate currents. Since these currents flow in the opposite direction from the current in the coil, the total magnetic field magnitude will be smaller. Since the inductance value can be defined as the ratio between total magnetic flux and coil current, the inductance of the coil will be reduced.

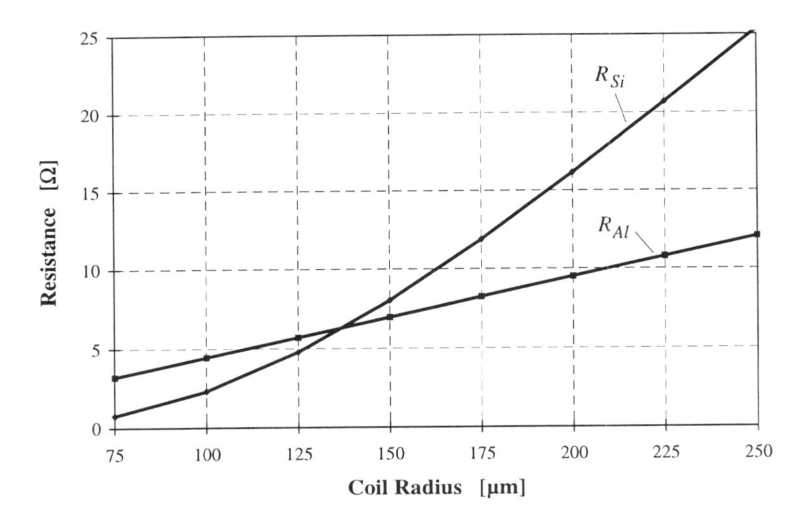

Figure 5.8. Simulation result for the resistive losses in the metal tracks and in the substrate for planar inductors on heavily doped substrates; inductor parameters are $n = 4$, $w = 9\mu m$ and $sp = 1\mu m$, frequency is $2GHz$ and substrate resistivity is $0.01\Omega cm$

For the largest coil with radius $r = 250\mu m$, the substrate currents decrease the value of L with approximately 10%.

Again, analytically analyzing this effect seems impossible. Therefore, the only way to date to make a safe design is the use of finite-element simulation. A general guideline is to limit the area of the coil. As can be seen from the simulation data of figure 5.8, the effect is far less severe for coils with a small radius r. This can be explained by the fact that for small coils, the magnetic field penetrates less deep into the substrate and therefore has less effect. Changing the distance from the substrate, as can be done for an inductor using the third or fourth metal level, does not effect this, since the change in oxide thickness will always be negligible compared to the depth of magnetic field penetration into the substrate.

5.2.2.3 Inductor Design Model. A drawback of both these designing methods, i.e. the calculation according to Greenhouse and the finite-element simulation, is the reverse way of designing necessary. One has to fix the geometry of the inductor, then determine its inductance value, and then see if it fits for the application in mind. A structured analog design method works in the other direction : first setting the inductance necessary by doing a proper circuit design and then sizing the inductor. This requires a simple model for the inductance value as a function of its parameters (number of turns, area, conductor width and spacing). A first example of such a model is given in [Voorm 1993].

Since this model is based on the Greenhouse formulas, it has the same limitations, i.e. all parasitics that appear at high frequencies and on conductive substrates are not taken into account. Using the simulation data available through the fast finite-element simulation strategy, an improved inductor model can be developed [Crols VLSI96, Crols 97]. This model is an extension of [Voorm 1993], since it now includes the effect of high-frequency magnetic fields on the series resistance. The losses in the conductive silicon substrate are not modeled, so it is only valid for inductors on a lowly doped substrate. The inductor is modeled with the lumped circuit model of figure 3.4, and the inductance L, the series resistance R_l, the parasitic capacitance C_p and its associated resistance R_c are modeled as a function of coil area, conductor width and spacing, area efficiency, frequency, etc. In the frequency range up to 3 GHz, the accuracy of the model is 10 % for the inductance value and 40 % for the series resistance. This model enables indeed a structured analog design approach with planar inductors, such as optimization of the technology, the geometry, and the circuit [Crols 97].

This model can be applied in a number of possible circuits, such as inductively tuned LNAs [Abidi ACD96, Rudel ISSCC97, Shaha ISSCC97], high-frequency bandpass output buffers [Abidi ISSCC97], input impedance matching circuits, etc. In these applications generally a rather low quality factor is needed, as the passband of the circuit must be wide enough to ensure proper operation over all possible process variations. The achieved accuracy in inductance and resistance value is then good enough. In oscillator design, however, high-Q inductors are needed and a 30-% accuracy in resistive losses is certainly unacceptable. So for VCO inductors, this structured analog design approach cannot be used and the optimum inductor must be chosen on a trial-and-error basis.

5.2.3 Hollow Coil Design Guidelines

From these discussions on the finite-element simulation results, we can remember four general design rules that have to be taken into account when sizing planar inductors on conductive silicon substrates. Because of the limited accuracy of the structured analog design model, quantitative results cannot be given, but the following are some qualitative guidelines :

- *Limit the width of the metal conductors* : This has two reasons. First, widening the metal conductors of an inductor with a fixed area will result in a smaller inductance value, as the inner turns become smaller and smaller and contribute less and less to the inductance. So to keep the inductance value constant, the coil area must be increased, resulting in longer conductor lengths and hence higher resistance. Secondly, there is the skin effect. This effect will cause a nonuniform current flow in the conductor, whereby the center of a wide conductor will not be used. So the high-frequency resistance increases once again.

- *Use minimum spacing between adjacent conductors* : This maximizes the positive mutual inductance, so enlarges the value of L, and allows for the largest center hole. The interwinding fringing capacitance will be negligible in most cases.

- *Do not fill the inductor up to the center* : This is a high-frequency magnetic field effect. Due to the generation of eddy currents at high frequencies, the innermost turns of the coil suffer from an enormous increase in resistance, while their contribution to the inductance value is already small. These inner turns deteriorate the overall quality factor, and an improvement in performance can be obtained by simply omitting them. So it is necessary to use a *hollow coil*.

- *Limit the area occupied by the coil* : This guideline is only valid for conductive substrates. At high frequencies, the magnetic field generated by the inductor induces currents in the substrate which cause extra resistive losses and a decrease in inductance value. If the substrate is lowly doped, this effect is generally negligible. For highly doped substrates, the only solution possible is to limit the size of the coils, as the magnetic field of small coils penetrates less deep into the substrate, and hence cause less losses.

Another study on integrated spiral oscillators [Long JSSC97], which uses transmission line models for the inductor segments, arrives basically at the same conclusions, except that the analysis technique applied there also cannot take the magnetic substrate losses into account.

These are of course only qualitative guidelines, and to evaluate their relative importance in a given application and technology the finite-element simulations must be repeated for a certain range of coil sizes. From the results of this simulation batch, the most optimal inductor for the specific application can be picked. This is demonstrated for two designs, which are described in the next two sections.

5.3 PLANAR-LC VCO DESIGN ON A HEAVILY DOPED SUBSTRATE

The technology available in this first demonstration design is the same standard 0.7-μm CMOS process that was used for the bonding wire VCO [Mietec C07]. It is in fact a worst-case technology for VCO planar inductor design for two reasons. First, there are only two metal layers, whereas some more modern technologies already offer three or more layers as a standard configuration. This limits the possible optimization of the metal losses. Secondly, the wafers used are of the epi-type and hence the substrate losses will play an important role in the optimization process. All the necessary data concerning the process are listed in table 5.2.

Substrate thickness	t_{subs}	625 μm
Substrate resistivity	ρ_{subs}	0.01 Ωcm
Epi-layer thickness	t_{epi}	10 μm
Epi-layer resistivity	ρ_{epi}	40 Ωcm
Field Oxide Thickness	t_{ox}	0.7 μm
Metal1 sheet resistance	$R_{\square,M1}$	50 $m\Omega/\square$
Metal2 sheet resistance	$R_{\square,M2}$	35 $m\Omega/\square$

Table 5.2. Physical parameters of the 0.7-μm CMOS process

Apart from the desired center frequency of 1.8 GHz, no strict specifications were given. A compromise must be sought between a phase noise as low as possible, at a low power consumption level and a reasonable tuning range. The design is done in two parts : first an optimal coil sizing is obtained, and then the active oscillator circuitry is developed.

5.3.1 Coil Geometry

For the design of the oscillator coil, several considerations must be made. Of course, the losses of the coil must be as low as possible for low noise and low power. One could use a tiny coil for this, or one with only one turn, since this will certainly have small resistance. But the power required for a stable oscillation is proportional to $R_{eff} \cdot (\omega_0 C)^2$ (see equation (3.45)). So using a too small inductance value will require a large capacitance to set the desired frequency, and hence a large power consumption will result. But the inductance value cannot be made very high either. Then the required capacitance value will be almost achieved with the parasitics of the coil and the amplifying transistors alone. This leaves no room for an extra tunable junction capacitance, so the tuning range will be small.

Due to the efficient finite-element simulation strategy of circular coils, we were able to evaluate a lot of possible coils. This allows to gain insight in the different trade-offs between coil radius r, conductor width w and number of turns n. For the relatively small inductance values that are aimed at, fringing capacitance between the several conductor tracks is negligible. So the conductor spacing sp should be chosen minimal, i.e. 1.5 μm in this technology. Both metal levels are used in parallel to decrease the resistance without widening the metal turns.

It is more efficient to use both metal layers in parallel, with a large number of vias in between them to make good contact, than placing two inductors with conductors of double width in series, one in metal1 and the other in metal2, as is done in [Razav

Radius	r	85 μm
Width	w	8.5 μm
Spacing	sp	1.5 μm
No. of Turns	n	4
Frequency	F	1.8 GHz
Inductance	L	3.2 nH
Metal Losses	R_{Al}	4.60 Ω
Silicon Losses	R_{Si}	1.75 Ω
Total Resistance	R_{eff}	6.35 Ω
Quality Factor	Q	5.7

Table 5.3. Optimized coil parameters for 0.7-μm VCO

ISSCC97]. They will both have approximately the same inductance and resistance value at low frequencies, but the latter one will suffer more from the skin effect. A slight disadvantage of the first structure is the connection of both inductor leads, since one of them is situated in the inside of the coil. Either the connection to this lead is done with a poly-line, at the expense of a large extra resistance, or it is done in metal1, while all crosses with this connection of the inductor turns are done in metal2 only. Because of the large polysilicon sheet resistance, certainly here because no silicided poly is available, the second option is taken.

After the proper setup, this simulation batch can be run overnight, and afterwards the inductor best fitted for the application in mind can be chosen. In this design, the inductor of our choice has parameters $r = 85\ \mu m$, $w = 8.5\ \mu m$, $sp = 1.5\ \mu m$ and $n = 4$. The other parameters of the inductor are given in table 5.3. With an inductance value of 3.2 nH, a total equivalent series resistance of 6.35 Ω is obtained. This leads to a quality factor of

$$Q = \frac{\omega L}{R_s} = \frac{2\pi \cdot 1.8GHz \cdot 3.2nH}{6.35\Omega} = 5.7 \tag{5.6}$$

These values are not the exact values given by the finite-element simulator, but were adjusted manually to account for the extra inductance and resistance caused by the connection leads. A more exact way than doing this manually would be to perform a 2-D planar simulation to investigate the effect of these connection leads. In a silicon layout, the circular coil can be adequately approximated by an octagonal shape. The size of this octagon is chosen such that the average radius is 85 μm.

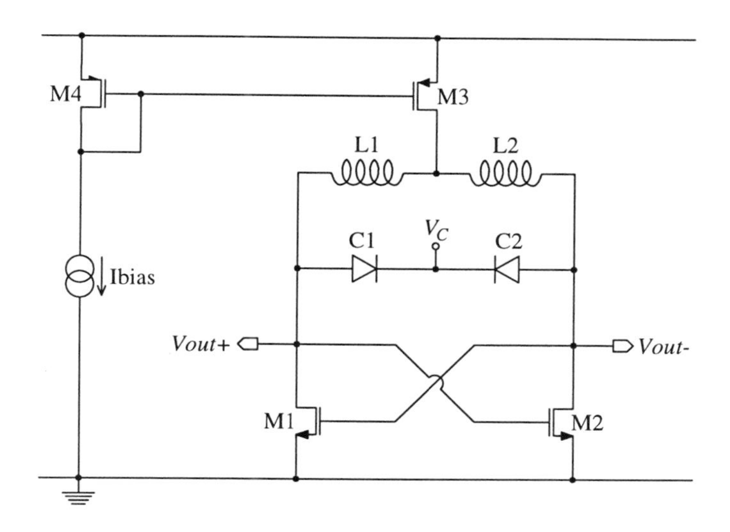

Figure 5.9. 0.7-μm Planar-LC oscillator circuit schematic

5.3.2 Amplifier Design

The oscillator circuit schematic is shown in figure 5.9. A differential configuration with two of the optimized hollow coils in series is used. This doubles the inductance value, and of course also doubles the value of the total series resistance. But since now the total carrier signal is 6 dB higher, and the effective resistance is only a factor of 2 larger, a 3-dB gain in phase noise is obtained. Since the effective capacitance value is halved, also required transconductance drops by a factor of two. As this G_M is realized with a differential configuration, the total power consumption is two times larger. A further power-noise trade-off with an enhanced LC-tank is not applied here.

The amplifier structure consists of two NMOS transistors $M1$ and $M2$ that are coupled in positive feedback to provide a negative resistance. This schematic allows for a very low power supply voltage, as there is no DC voltage drop across the inductors. The minimum power supply is $V_{DSat,M3} + V_{GS,M1}$.

With an inductance value of 3.2 nH, the total capacitance on each node must be 2.4 pF to obtain an oscillation frequency of 1.8 GHz. The LC-tank's effective resistance will be, including the parasitic resistances of the capacitors, approximately 15 Ω. So the required negative conductance provided by the transistors must be $15\Omega \times (2\pi \times 1.8GHz \times 1.2pF)^2 = 3mS$. Using a safety factor of two, each of the amplifying transistors must have a transconductance g_m of 12 mS.

The capacitor of the LC-tank is formed by the inductor's parasitic capacitance to the substrate, the drain-bulk, gate-drain and gate-source capacitances of the NMOS transistors, and a tunable p+/n-well junction capacitor. In order to achieve a large tuning range, this last contribution must be as large as possible.

The parasitic capacitance of the coil is only 0.2 pF because of the small dimensions used. This leaves 2.2 pF to be divided between the transistors' parasitics and the tuning capacitor. For a minimum gate length transistor (i.e. 0.7 μm), the sum of drain-bulk, gate-drain and gate-source capacitance in this technology approximately equals 3 fF per μm gate length. Very small values of $V_{GS} - V_T$ for $M1$ and $M2$ would yield a large transconductance-to-current ratio and hence a small power consumption. However, in that case the transistor sizes and hence its parasitic capacitances become too large and the tuning range will be small. In this design a $V_{GS} - V_T$ value of 0.3 V has been chosen. This yields a width of 400 μm for transistors $M1$ and $M2$ and a total current drain of 4 mA. With the power supply of 1.5 V, this results in only 6 mW total power consumption. With an oscillation amplitude of approximately 1.1 $V_{peak,diff}$, the expected phase noise can be calculated to be :

$$\begin{aligned} \mathcal{L}\{600kHz\} &= \frac{kT \cdot 15\Omega \cdot [1+2] \cdot \left(\frac{1.8GHz}{600kHz}\right)^2}{\frac{1.1V^2}{2}} \\ &= 2.75 \cdot 10^{-12} \\ &= -115.6\,dBc/Hz \end{aligned} \tag{5.7}$$

The 1-pF tunable junction capacitor is made as a p+ active area in an n-well. Its capacitance can be tuned with the control voltage V_C, which controls the bias voltage of the n-well. Since this n-well is a common-mode node, its parasitic capacitance to the substrate is not important. Of course, the layout of this junction capacitor is very important. Care must be taken to limit the series resistance, and to keep the symmetry that guarantees the common-mode nature of the n-well. A small part of the junction capacitor layout is shown in figure 5.10. All active area stripes have minimal width, and the stripes of the two capacitors $C1$ and $C2$ are interdigitated, and interrupted by contacts to the common N-well.

5.3.3 Measurement Results

A microphotograph of the VCO is shown in figure 5.11. The die measures 750 $\times$ 750 μm^2. The two oscillator coils are situated on the top. The other coils are used as loads for the measurement output buffers. These output buffers are simple common-source transistors with an inductive load to compensate for the bonding pad parasitics. The tunable capacitors and the amplifying transistors are placed in the middle of the die.

The free-running oscillation frequency is 1.81 GHz, which is only 4% off from the predicted value of 1.88 GHz. As said earlier, performing an analysis of the effects of

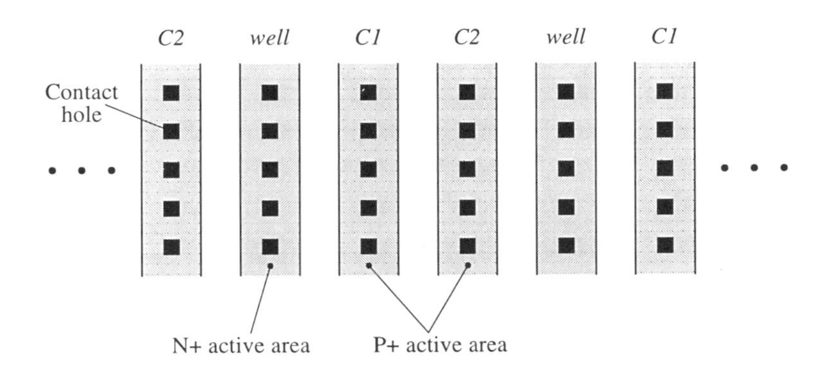

Figure 5.10. Low-resistance differential junction capacitor layout scheme

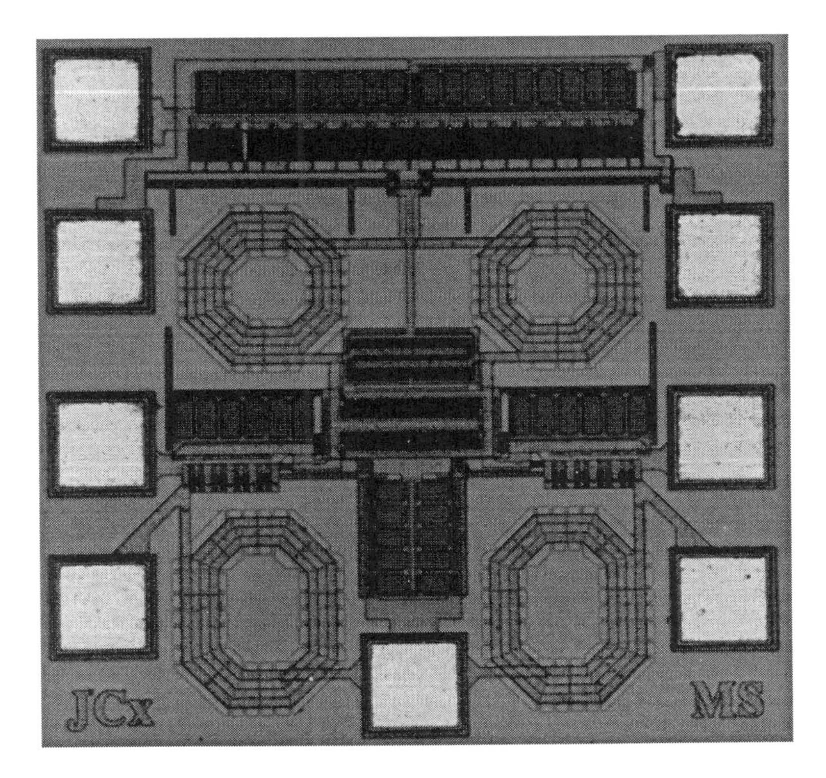

Figure 5.11. IC microphotograph of $0.7\text{-}\mu m$ planar-LC VCO

the coil connection leads with a planar 2-D or even a full 3-D simulator might have resulted in an even better prediction of the center frequency. The oscillator operates from a 1.5-V power supply and consumes only 4 mA. The measured output spectrum is shown in figure 5.12(a). The center frequency is 1.81 GHz and the resolution bandwidth is 10 kHz. A logarithmic plot of the phase noise measured with a spectrum analyzer is given in figure 5.12(b). The resulting phase noise is -116 dBc/Hz at 600 kHz offset, which agrees very well with the theoretical value of equation (5.7). For offset frequencies larger than 100 kHz, the phase noise decreases with a slope of 20 dB/dec. This shows that $1/f$ noise only becomes important below 100 kHz offset.

The frequency tuning with control voltage V_C is shown in figure 5.13. The voltage can be put as low as 0.5 V, to a point where the p+/n-well junction becomes slightly forward biased. At that time the phase noise is approximately 3 dB worse than the result shown in figure 5.12. Two facts explain this. First, the total capacitance of the LC-tank becomes larger, in order to lower the oscillation frequency. This increases the transconductance required to maintain oscillation. Since there is no amplitude control implemented in this VCO, the negative resistance implemented by the transistors $M1$ and $M2$ remains the same. This causes a 1-dB decrease in oscillation amplitude and thus in phase noise. The other 2 dB loss is probably caused by increasing losses in the almost-forward-biased junction capacitors, or, as will be seen in the next design, by the increase in upconverted baseband noise due to the larger nonlinearities.

5.4 PLANAR-LC VCO DESIGN ON A LOWLY DOPED SUBSTRATE

For this second design that demonstrates the capabilities of planar integrated silicon inductors when adequately optimized, a deep submicron 0.4-μm standard CMOS technology is available. It is better suited for high-frequency analog design than the previous technology in the sense that here bulk wafers are used instead of epi-wafers. The number of metal levels is still limited to two. But due to the absence of large silicon losses in this lowly-doped (and thus highly resistive) substrate, another improvement in phase noise with respect to the previous design will be possible. A summary of the most important process parameters is given in table 5.4.

Actually, two oscillators were made. Because of the possible improvement by the absence of substrate losses, it will now be feasible to make a VCO design that achieves the required phase noise specs of the GSM system, and of its higher-frequency relative, DCS-1800. The required specs were already calculated in section 2.2 to be -121 and -119 dBc/Hz, respectively, at an offset frequency of 600 kHz from the carrier. Again, the design path can be split up into two parts, i.e. optimization of the inductor coil and sizing of the amplifier.

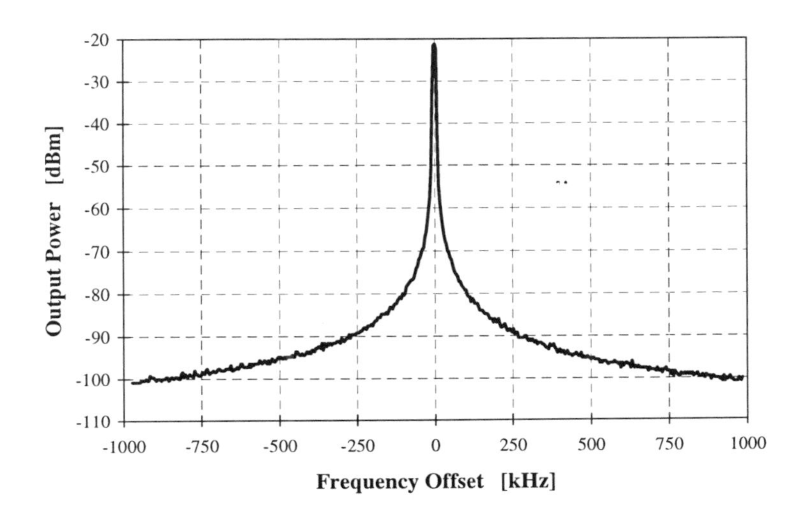

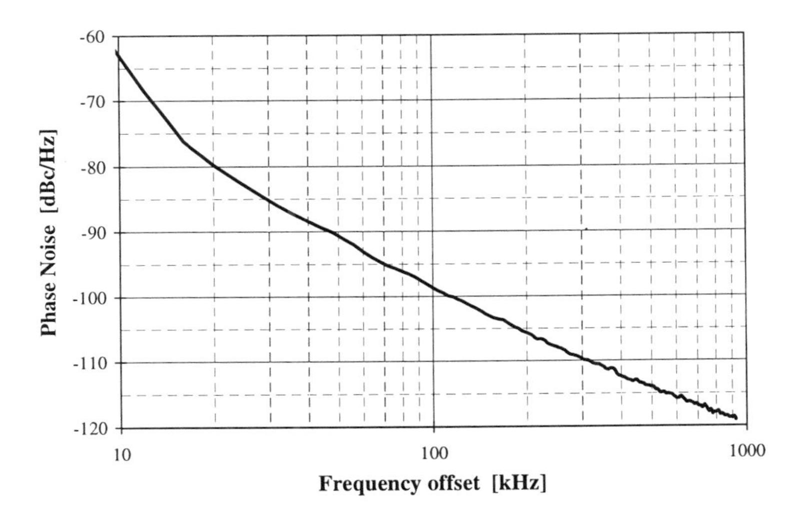

Figure 5.12. Measured phase noise of the $0.7\text{-}\mu m$ planar-LC VCO : (a) Output spectrum for a carrier frequency of $1.81\ GHz$ and resolution bandwidth of $10\ kHz$; (b) Logarithmic plot

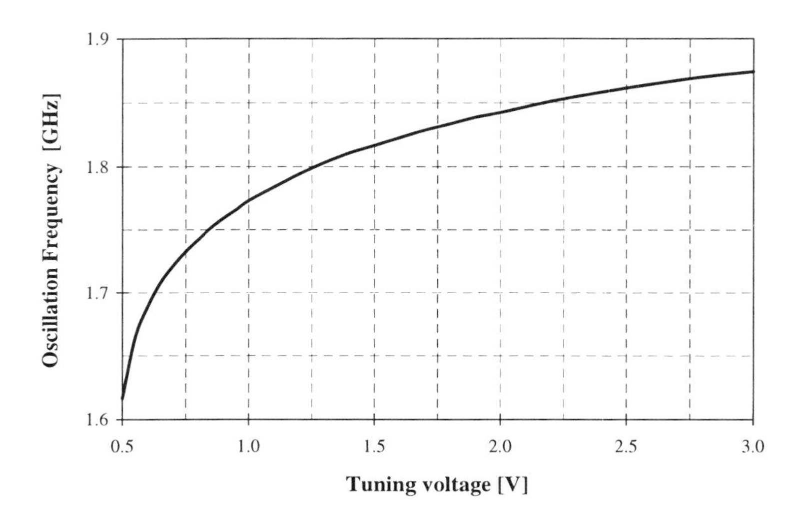

Figure 5.13. Measured tuning characteristic of the $0.7\text{-}\mu m$ planar-LC VCO

Substrate thickness	t_{subs}	$625\ \mu m$
Substrate resistivity	ρ_{subs}	$5\ \Omega cm$
Epi-layer thickness	t_{epi}	No epi-layer
Epi-layer resistivity	ρ_{epi}	used !
Field Oxide Thickness	t_{ox}	$0.7\ \mu m$
Metal1 sheet resistance	$R_{\square,M1}$	$50\ m\Omega/\square$
Metal2 sheet resistance	$R_{\square,M2}$	$35\ m\Omega/\square$

Table 5.4. Physical parameters of the $0.4\text{-}\mu m$ CMOS process

5.4.1 Coil Geometry

As the substrate type is lowly doped, there is no longer a strong limit on the area occupied by the coil. So only the first three guidelines mentioned is section 5.2.3 are valid in this design. We will see in the optimization that this generally leads to larger values of L for the optimal coil. This is not always an advantage, as large inductance values lead to a smaller capacitance value and hence less tuning range, if an important part of the capacitance is already accounted for by the coil and transistor parasitics. Using two coils in series as was done in the previous design will result in

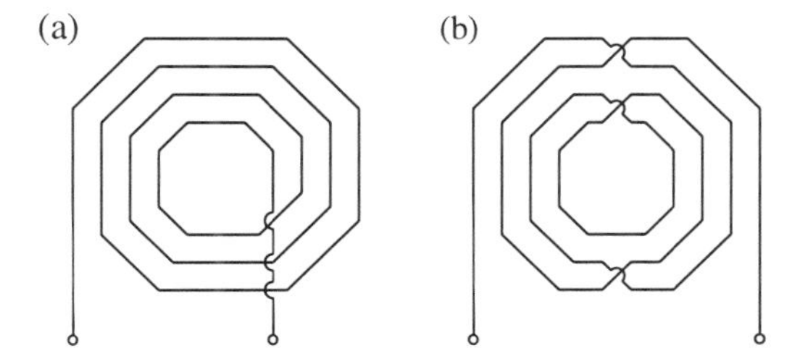

Figure 5.14. Planar inductor coil shape : (a) standard octagonal; (b) symmetrical

this constraint. But since the differential oscillator configuration has the intrinsic 3-dB phase noise gain with respect to single-ended implementations, there is a need for a perfectly balanced inductor shape. As shown in figure 5.14(a), a normal octagonal coil shape can never be perfectly symmetrical.

The newly developed coil shape is shown in figure 5.14(b). Instead of drawing a normal spiral track, and then making the connection to the inner turn in a different metalization level, the metal track now moves inside every half turn, until the center is reached, and then moves outside again every half turn. Both the Metal1 and Metal2 routing levels are used, except for the cross sections which use of course only one metal level for each direction. The number of cross sections is equal for both shapes, so this new shape has no effect on the DC series resistance. This coil is completely symmetrical and can thus be used as a differential floating inductor. A disadvantage of this coil is the fact that no common mode point is available externally, as was the case for the series connection of two standard octagonal coils of figure 5.9. The common mode point is the innermost turn and is difficult to reach from the outside, because both metalization levels are already used. But looking at figure 5.9, we can see that in this circuit a resistance situated in series with $M3$ will have little effect on the oscillator circuit, as this operates completely differential. It is therefore allowable to route the common mode connection in e.g. the polysilicon level, which has a large sheet resistance.

Table 5.5 lists the results of batch finite-element simulation run that was used to determine the optimum coil sizes of the 900-MHz VCO. A number of coils were simulated, with radii r ranging from 120 to 180 μm, conductor widths w from 6 to 15 μm and 4 to 6 turns. The conductor spacing sp was always equal to the minimal value of 1 μm and all simulations were of course done at a frequency of 0.9 GHz. The table lists the inductance value (L), the contribution to the total resistance of the metal losses (R_{Al}) and the substrate losses (R_{Si}), as well as the overall quality factor (Q). Because of the lowly-doped wafers, the substrate losses are always negligible. Furthermore, the capacitance value required to set the frequency at 0.9 GHz is listed (C)

and the necessary transconductance (G_M). The last column gives a preliminary estimation of the achievable phase noise at 600 kHz offset, assuming an amplifier noise contribution factor A of 3, and a carrier amplitude of 1 $V_{rms,diff}$.

The inductor of our choice is no. 9, and is marked in bold font in the table. The coil parameters are $r = 120\ \mu m$, $n = 4$, $w = 7.5\ \mu m$ and $sp = 1\ \mu m$. As can be seen, this is not the inductor with the highest quality factor, but it is the inductor that is best fitted for the application. The highest Q-values are achieved by the coil with the largest radius, but those generally have a larger parasitic capacitance, which will fill in a larger part of the already smaller required total tank capacitance. This will limit the tuning range.

Another aspect that has not been discussed yet, is the radiation of the VCO signal to the other parts of a fully integrated RF front-end. This will cause the well-known LO-to-RF leakage in up- and downconversion mixers, which results in unwanted spurious signals in their output. Quantitative analysis of this effect is of course not possible, but qualitatively it is easily understood that the magnetic field of large coils will radiate stronger to nearby circuits and will hence cause more LO-to-RF leakage. This has also been a motivation to choose a coil with a limited radius.

The initial circuit parameters of the coil are $L = 9.43\ nH$ and $R_l = 9.40\ \Omega$. Again, these numbers must be adjusted for the additional inductance and resistance of the connection leads. The corrected values chosen here, which might be an overestimation, are $L = 10.3\ nH$ and $R_l = 10.5\ \Omega$. The left-hand side column of table 5.6 lists the final coil parameters.

One aspect of the newly developed symmetrical coil shape that has not been dealt with, is the fringing capacitance between the segments. In a standard spiral coil with a certain number of turns, the voltage between the outer track and the one next to that is not that big. So the voltage swing across the fringing capacitance is limited, and it will have little effect on the total parasitic capacitance of the coil. But in the symmetrical coil almost the complete voltage swing stands across these adjacent segments. The fringing capacitance will therefore become more important and its effect must be incorporated in the simulations. Since the lumped equivalent circuit model of figure 3.4 is no longer sufficient, a more accurate model is presented in figure 5.15.

The coil is split up in segments that each represent one half of a turn. Those segments are now modeled as seven-terminal devices, as shown in figure 5.15(b). The first two terminals are the normal ones, i.e. both ends of the useful inductor, and are indicated with a round shape ($\circ$). The inductance value L is split up in four parts, the series resistance R_l in three parts. At the two intermediate nodes that appear in this way the parasitic capacitances are connected. There is of course the capacitance to the substrate, C_p, and its associated resistance R_c. They are connected to the third terminal of the segment, i.e. the substrate connection. The four other terminals are used to connect the fringing capacitances C_f to the adjacent segments, and are indicated with

	r [μm]	n [-]	w [μm]	L [nH]	R_{Al} [Ω]	R_{Si} [Ω]	Q [-]	C [pF]	G_M [mS]	$\mathcal{L}\{600kHz\}$ [dBc/Hz]
1	120	4	6	6.17	8.55	0.012	4.07	5.07	7.04	-125.0
2	120	4	9	5.00	5.55	0.010	5.09	6.25	6.95	-126.9
3	120	4	12	4.10	4.11	0.008	5.63	7.63	7.66	-128.2
4	120	4	15	3.36	3.28	0.007	5.78	9.31	9.11	-129.2
5	120	5	6	8.52	10.37	0.017	4.64	3.67	4.47	-124.2
6	120	5	9	6.64	6.65	0.014	5.63	4.71	4.73	-126.1
7	120	5	12	5.20	4.88	0.011	6.01	6.01	5.66	-127.4
8	120	6	6	10.91	12.05	0.022	5.11	2.87	3.17	-123.5
9	**120**	**6**	**7.5**	**9.43**	**9.38**	**0.019**	**5.67**	**3.32**	**3.31**	**-124.6**
10	120	6	9	8.15	7.64	0.017	6.02	3.84	3.60	-125.5
11	120	6	12	6.07	5.56	0.013	6.16	5.15	4.73	-126.9
12	150	4	6	8.58	10.95	0.021	4.42	3.64	4.66	-123.9
13	150	4	9	7.18	7.19	0.018	5.63	4.36	4.38	-125.8
14	150	4	12	6.08	5.36	0.016	6.39	5.14	4.55	-127.0
15	150	4	15	5.17	4.31	0.014	6.78	6.04	5.04	-128.0
16	150	5	6	12.06	13.37	0.030	5.09	2.59	2.88	-123.1
17	150	5	9	9.78	8.70	0.025	6.34	3.20	2.85	-124.9
18	150	5	12	8.02	6.46	0.021	7.00	3.90	3.15	-126.2
19	150	5	15	6.58	5.18	0.018	7.16	4.75	3.75	-127.2
20	150	6	6	15.72	15.66	0.040	5.67	1.99	1.99	-122.4
21	150	6	9	12.36	10.10	0.033	6.90	2.53	2.07	-124.3
22	150	6	12	9.78	7.45	0.026	7.40	3.20	2.44	-125.6
23	150	6	15	7.70	5.97	0.021	7.28	4.06	3.15	-126.6
24	180	4	6	11.13	13.35	0.032	4.70	2.81	3.38	-123.1
25	180	4	9	9.48	8.82	0.029	6.06	3.30	3.08	-124.9
26	180	4	12	8.20	6.62	0.026	6.98	3.81	3.09	-126.1
27	180	4	15	7.14	5.34	0.023	7.54	4.38	3.29	-127.0
28	180	5	6	15.81	16.38	0.047	5.44	1.98	2.06	-122.2
29	180	5	9	13.15	10.75	0.042	6.89	2.38	1.95	-124.0
30	180	5	12	11.09	8.03	0.036	7.77	2.82	2.05	-125.3
31	180	5	15	9.39	6.48	0.031	8.16	3.33	2.31	-126.2
32	180	6	6	20.85	19.27	0.064	6.10	1.50	1.39	-121.5
33	180	6	9	16.91	12.55	0.055	7.59	1.85	1.38	-123.3
34	180	6	12	13.88	9.34	0.046	8.36	2.25	1.52	-124.6
35	180	6	15	11.40	7.52	0.038	8.53	2.74	1.82	-125.5

Table 5.5. Finite-element batch simulation results for 900-MHz VCO inductor in 0.4-μm technology

Application :		GSM	DCS-1800	
Radius	r	120	120	$[\mu m]$
Width	w	7.5	15	$[\mu m]$
Spacing	sp	1	1	$[\mu m]$
No. of Turns	n	6	3	[-]
Frequency	F	0.9	1.8	$[GHz]$
Inductance	L	10.3	2.8	$[nH]$
Total Resistance	R_{eff}	10.5	3.7	$[\Omega]$
Quality Factor	Q	5.6	8.6	[-]

Table 5.6. Optimized coil parameters for both 0.4-μm VCOs

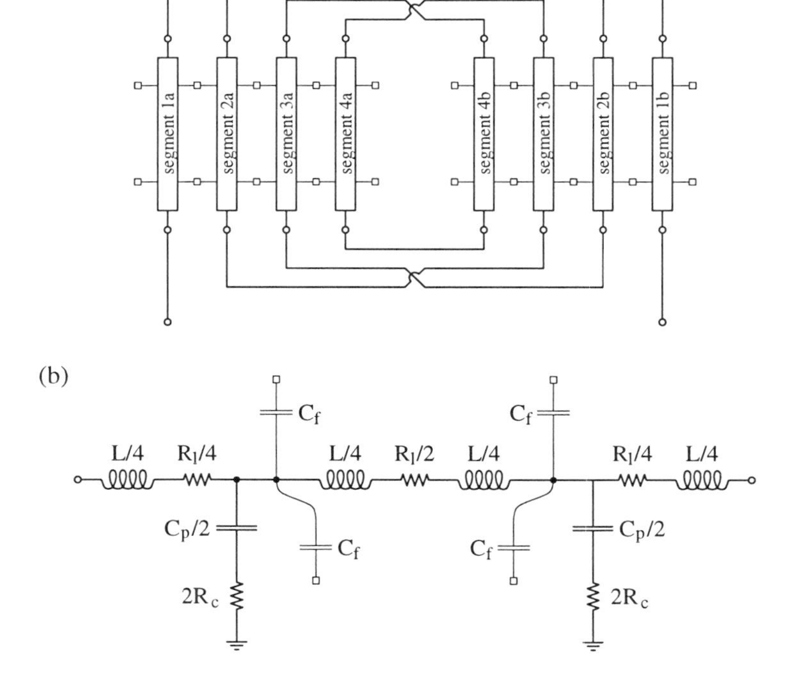

Figure 5.15. (a) Equivalent circuit model for symmetrical octagonal coil; (b) Model of a single segment

the square shape (□). The value of the fringing capacitances has been obtained with finite-element simulations and is approximately 0.15 fF per μm length.

The 1.8-GHz coil was designed in parallel with the same procedure. Its parameters were already listed in the right-hand side column of table 5.6. The DCS-1800 phase noise spec is 2 dB less severe than the GSM spec, but due to the double carrier frequency, it is actually 4 dB tougher. Both coils have the same radius, but the DCS-1800 coil has half the number of turns, and the metal tracks are twice as wide. The series resistance should be roughly four times lower, but this is not exactly true because of the high-frequency magnetic field effects. The finite-element simulations have been used to monitor and limit this increase in resistance.

5.4.2 Amplifier Design

The oscillator circuit schematic is shown in figure 5.16. The negative resistance is formed by the double cross-connection of an NMOS ($M1$-$M2$) and a PMOS ($M3$-$M4$) differential pair in positive feedback. In a 3-V system, this allows to use the supply current twice for amplification. A circuit such as the one of figure 5.9 runs at half the supply voltage, but at twice the current drain. A drawback of this configuration is of course the larger parasitic capacitances of the PMOS transistors with respect to an NMOS-only implementation. This will result in a slightly smaller available tuning range as the fixed part of the tank capacitance has increased. An advantage is that there is no need for a connection to the common-mode point of the inductor, which would be a problem for the new symmetrical coil shape.

The $V_{GS} - V_T$ and the g_m of the amplifying transistors must be chosen correctly in order to achieve a good compromise between power consumption, phase noise and tuning range. Again, a low value of $V_{GS} - V_T$ gives a good transconductance-to-current ratio and hence a low power consumption, but results in large transistors and a small the tuning range. In both designs a value of 0.33 V is chosen. This is slightly larger than the value chosen in the 0.7-μm design, in order to compensate a little bit for the increase in parasitic capacitance due to the larger PMOS transistors.

For the GSM VCO, the effective resistance of the LC-tank equals 11.7 Ω. The extra 1.4 Ω with respect to the data in table 5.6 comes from the extra losses in the junction capacitor and the transistors. The required negative transconductance of the amplifier must then be at least equal to

$$G_{M,GSM} = \frac{R_{eff}}{(\omega_0 L)^2} = \frac{11.7\Omega}{(2\pi \cdot 900MHz \times 10.3nH)^2} = 3.5\ mS \qquad (5.8)$$

The safety factor in the transconductance value must be large enough to ensure proper start-up of the oscillator, and is chosen to be 2.5. Each transistor $M1$-$M4$ must than have a transconductance of $3.5mS \times 2.5 = 8.75mS$. With the $V_{GS} - V_T$ value of

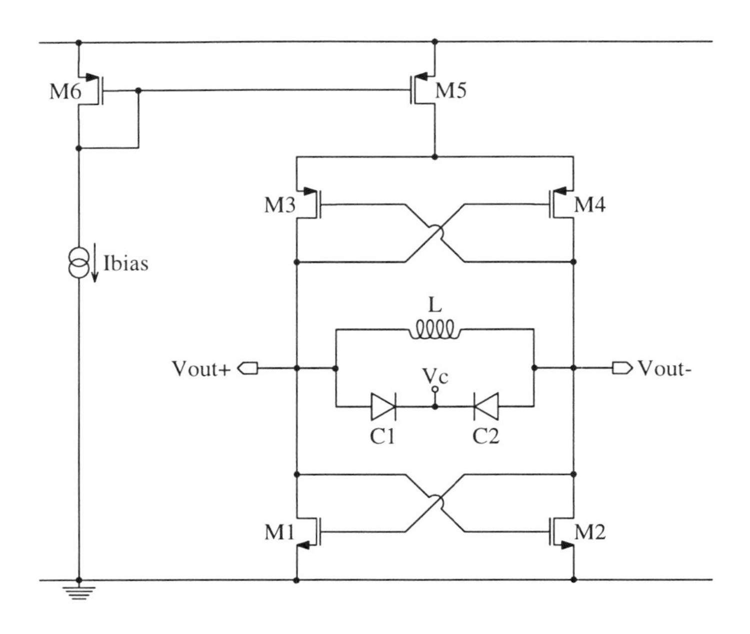

Figure 5.16. 0.4-μm Planar-LC oscillator circuit schematic

$0.33\ V$, the total current consumption is

$$I_{GSM} = 2 \cdot I_{M1} = 2 \cdot \frac{g_{m,M1} \times (V_{GS} - V_T)_{M1}}{2} = 8.75mS \times 0.33V \approx 3mA \quad (5.9)$$

The NMOS transistors $M1$ and $M2$ are 80 μm wide, and the PMOS transistors are approximately three times larger, or 255 μm. This combination gives a single-ended oscillation amplitude of 1.3 V_{ptp}. The expected phase noise at 600 kHz offset then equals

$$\begin{aligned} \mathcal{L}\{600kHz\} &= \frac{kT \cdot 11.7\Omega \cdot [1 + 2.5] \cdot \left(\frac{900MHz}{600kHz}\right)^2}{\frac{1.3V^2}{2}} \\ &= 4.47 \cdot 10^{-13} \\ &= -123.5dBc/Hz \end{aligned} \quad (5.10)$$

which is 2.5 dB better than required by the GSM specs.

A similar calculation can be made for the DCS-1800 VCO. The effective resistance equals 4.6 Ω, the required amplifier transconductance is 4.5 mS, which results in a power consumption of 3.7 mA. Transistor sizes are 105 μm for $M1$-2, and 340 μm

for $M3$-4. The expected phase noise at 600 kHz offset equals -121.5 dBc/Hz, which is also 2.5 dB better than the system specs.

As can be seen in figure 5.16, the tuning of the VCO is again done through the variable capacitance of two P+/N-well junctions with the control voltage V_C. The maximum frequency will be obtained with the highest tuning voltage of 3 V. The varicap is sized to set this highest frequency at 1.02 GHz for the GSM VCO and 1.98 GHz for the DCS-1800 VCO. By lowering the control voltage down to 0.6 V, when the diodes become slightly forward biased, the lowest obtainable frequencies should be 0.82 GHz and 1.60 GHz, respectively. Of course, sizing of the tuning capacitor, determining its parasitic resistance, sizing of the amplifier transistors and determining their parasitic capacitances must be done in parallel in a few iterations. The junction capacitor layout is again done as shown in figure 5.10 for low series resistance.

5.4.3 Measurement Results

Both designs were laid out and processed in the 0.4-μm CMOS process. A simple resistor-loaded common-source NMOS transistor was used as an output buffer to drive the 50-Ω measurement system. First, the measured results of the 900-MHz design are presented.

5.4.3.1 900-MHz Design. A microphotograph of the GSM VCO is shown in figure 5.17. The die measures 725 $\times$ 825 μm^2. The oscillator coil can clearly be seen on the top. Underneath that, the tuning capacitors are placed, followed by the PMOS transistors $M3$-$M4$ and the NMOS transistors $M1$-$M2$. The core oscillator circuit only occupies 250 $\times$ 550 μm^2, half of which must be contributed to the inductor coil. The rest of the die is occupied by the output buffers, power supply decoupling capacitance, and the bonding pads.

The measured free-running oscillation frequency is 1.03 GHz, which is only 1% off from the predicted value of 1.02 GHz. This extremely good result shows that the finite-element simulations and the segmented coil circuit model are accurate enough for first-time-right VCO designs. The power consumption is 9 mW from a 3-V supply voltage. The measured output spectrum with a resolution bandwidth of 10 kHz is shown in figure 5.18(a). A logarithmic plot of the phase noise is given in figure 5.18(b). The resulting phase noise is -108 dBc/Hz at 100 kHz offset. There is only a small region in this spectrum analyzer measurement where the phase noise decreases with a 20 dB/dec slope. For smaller offset frequencies, the VCO enters its $\Delta\omega^{-3}$ region, whereas for larger offset frequencies the signal power drops below the noise floor of the measurement setup. However, if we extrapolate the measured results assuming a 20 dB/dec slope up to 600 kHz offset, the phase noise equals -123.5 dBc/Hz, which is exactly the number predicted by (5.10).

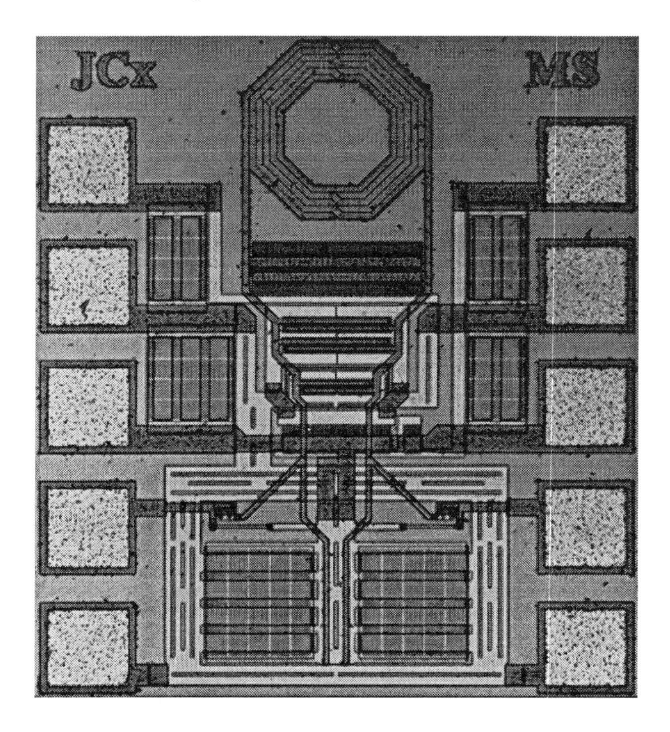

Figure 5.17. IC microphotograph of the planar-LC GSM VCO

This VCO has also been measured with the delay line technique as depicted in figure 4.23. That way a distinction between AM and PM noise can be made, and due to the larger dynamic range also phase noise levels up to -160 dBc/Hz can be measured. The resulting phase noise plot is shown in figure 5.19. Again, we notice that due to the absense of AM noise this result is approximately 3 dB lower than the spectrum analyzer result. The measured value at 100 kHz offset is -111 dBc/Hz. Also, the phase noise keeps falling off for higher offset frequencies with a slope of approximately -20 dB/dec. This validates the previous extrapolation to 600 kHz. The measured value at 600 kHz offset is now -125 dBc/Hz. The measured phase noise level at 3 MHz offset is -143 dBc/Hz.

The frequency tuning with control voltage V_C is shown in figure 5.20. The usable frequency range is $0.84 - 1.03$ GHz, which shows that tuning is possible over a range as wide as 20 %.

An unexpected problem that has been encountered during the measurements of this VCO is the significant increase of the $\Delta\omega^{-3}$ phase noise level at low tuning voltages. This noise originates from low-frequency $1/f$ noise that is upconverted to the carrier

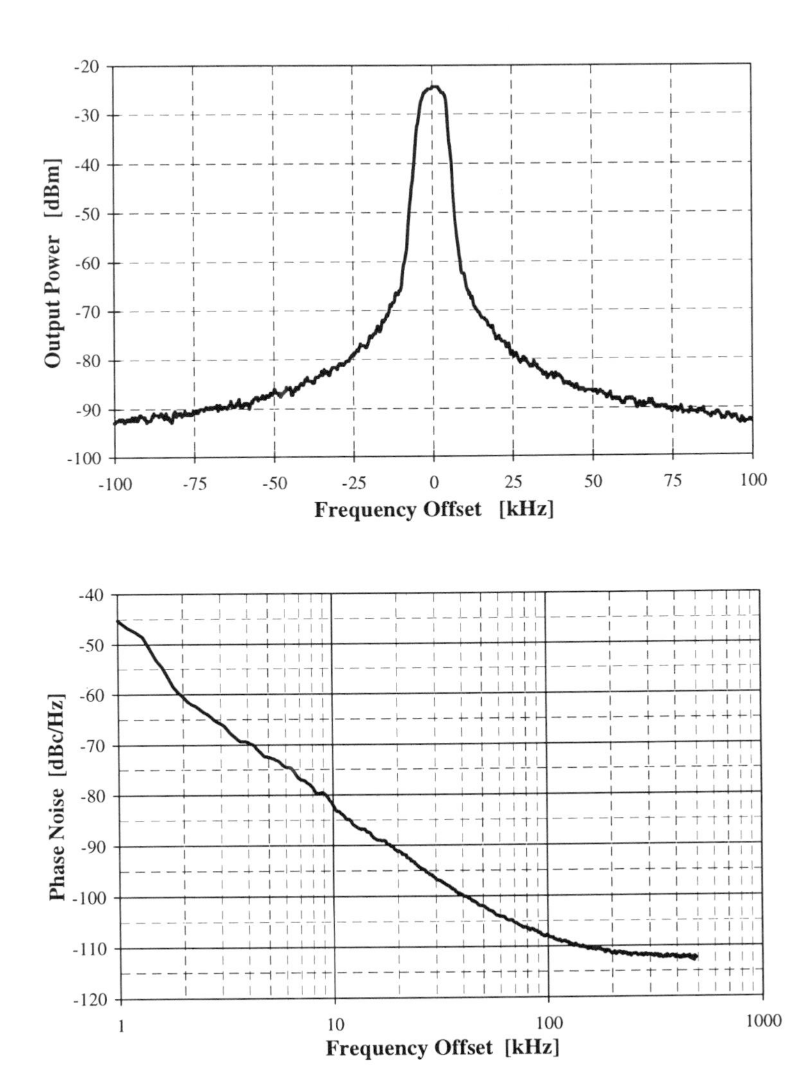

Figure 5.18. Measured planar-LC GSM VCO phase noise : (a) Output spectrum for a carrier frequency of $1.03\ GHz$ and resolution bandwidth of $10\ kHz$; (b) Logarithmic plot

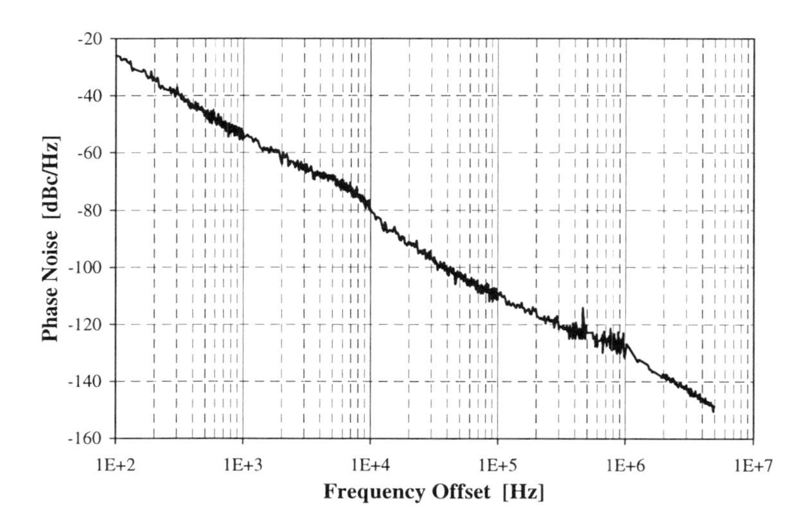

Figure 5.19. Planar-LC GSM VCO phase noise measured with the delay line method

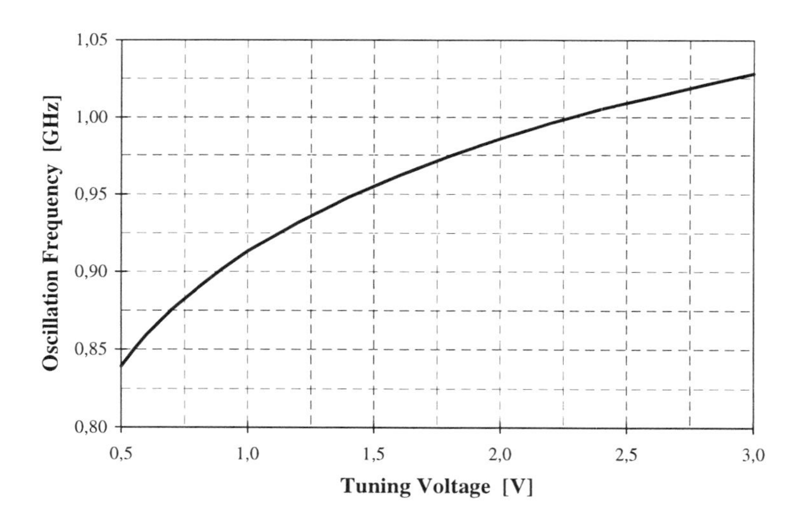

Figure 5.20. Measured planar-LC GSM VCO tuning characteristic

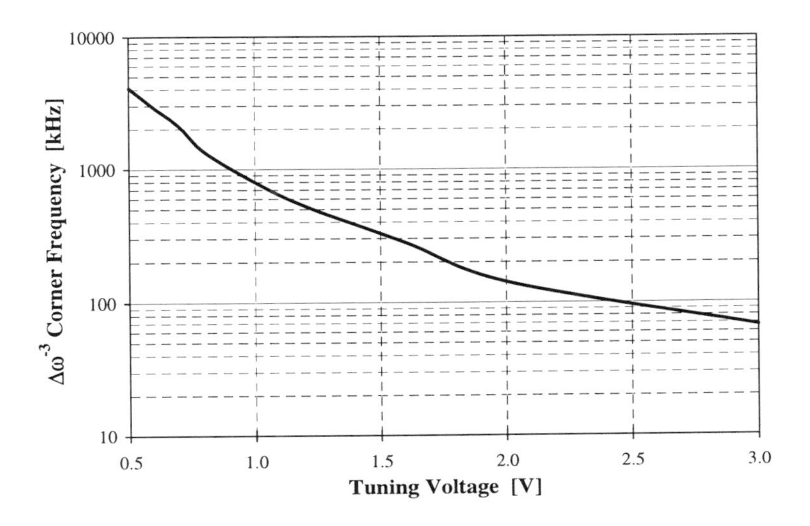

Figure 5.21. $\Delta\omega^{-3}$ Corner frequency vs. tuning voltage of the planar-LC GSM VCO

frequency by non-linearities in the circuit. As is predicted by [?], the magnitude of this $\Delta\omega^{-3}$ phase noise is determined by the symmetry properties of the oscillation waveform. As we use a good-quality LC-tank in a differential configuration, the VCO signal is a very good approximation of a sine wave and the phase noise due to $1/f$ transistor noise is limited. That is why the corner frequency between the $\Delta\omega^{-3}$ and the $\Delta\omega^{-2}$ region is below 100 kHz, in contrast to the much larger values reported for e.g. ring oscillators. However, as we reduce the reverse bias across the junction diode capacitors, the nonlinearities and asymmetries in the oscillator increase, thereby causing an increase in $\Delta\omega^{-3}$ phase noise.

In the previous design, its level always stayed negligible at the offset frequencies of interest, thereby indicating that $1/f$ noise is of no importance in these oscillators. But as the transistor sizes in this technology have a minimal gate length of 0.4 μm, the transistor areas have become much smaller, and hence the $1/f$ noise is larger. Also, the Spice KF factor that models the magnitude of the $1/f$ noise [HSpice96] is twice as large as for the 0.7-μm technology, further deteriorating the performance. This could have been remedied by the use of the (larger) PMOS transistors, which generally have a $1/f$ noise level an order of magnitude lower than NMOS transistors. This is true if the PMOS transistors are of the buried-channel type, but here they have a channel at the surface of the silicon. The surface scattering mechanism lead to a Spice KF factor twice as large as the NMOS transistors. The combination of these effects leads to a shift of the $\Delta\omega^{-3}$ corner frequency towards higher frequencies for low tuning voltages, as is shown in figure 5.21. For low frequencies the corner frequency is larger than 600 kHz and therefore corrupts the GSM phase noise spec.

5.4.3.2 1.8-GHz Design. The layout of the DCS-1800 VCO is, except for the coil and some minor changes in the transistor sizes, a perfect match of the 900-MHz design. The measured free-running oscillation frequency is 1.99 GHz, which is again only 0.5% off from the predicted value of 1.98 GHz. The power consumption is as expected, i.e. 11 mW from a 3-V supply voltage. The measured output spectrum with a resolution bandwidth of 10 kHz is shown in figure 5.22(a). A logarithmic plot of the phase noise is given in figure 5.22(b). The resulting phase noise is -113 dBc/Hz at 200 kHz offset. Extrapolating this to 600 kHz results in -122.5 dBc/Hz, which is 1 dB better than predicted. The frequency tuning is shown in figure 5.23. A 20-% range from 1.62 to 1.99 GHz can be covered. This oscillator suffers from the same problem with $1/f$ noise as the 900-MHz VCO.

5.5 CONCLUSIONS

After the state-of-the-art bonding wire inductors in the previous chapter, this chapter has dealt with the more "normal" type of integrated inductor, i.e. the planar inductor. If the phase noise spec allows it, they are preferred over bonding wires because of their smaller area, better accuracy and more confident manufacturability. Up to now, planar inductors were mostly modeled by the first-order calculation method of [Green TPHP74]. However, a lot of parasitics that appear at high frequencies and on conductive substrates are not taken into account.

The only way to incorporate all magnetic field effects is by finite-element simulation. As an alternative to a full 3-D model, we have proposed to exploit the symmetry around the vertical axis of circular instead of square inductors. This very fast simulation method allows to analyze quickly a lot of coil sizes, and has led to the following guidelines :

- *Limit the width of the metal conductors* : This is due to a more important increase in the high-frequency resistance because of the skin-effect.

- *Use minimum spacing in between the conductors* : This will maximize the inductance value, and the fringing capacitance is generally negligible.

- *Do not fill the inductor up to the center* : The innermost tracks contribute only a small amount of inductance, but due to eddy-currents generated by the magnetic field of the outer turns, their high-frequency resistance increases enormously. It is best if they are simply omitted. This leads to the *hollow coil* shape.

- *Limit the area occupied by the coil* : This way the magnetic field penetrates less deep into the substrate, and the associated substrate losses will be smaller. This

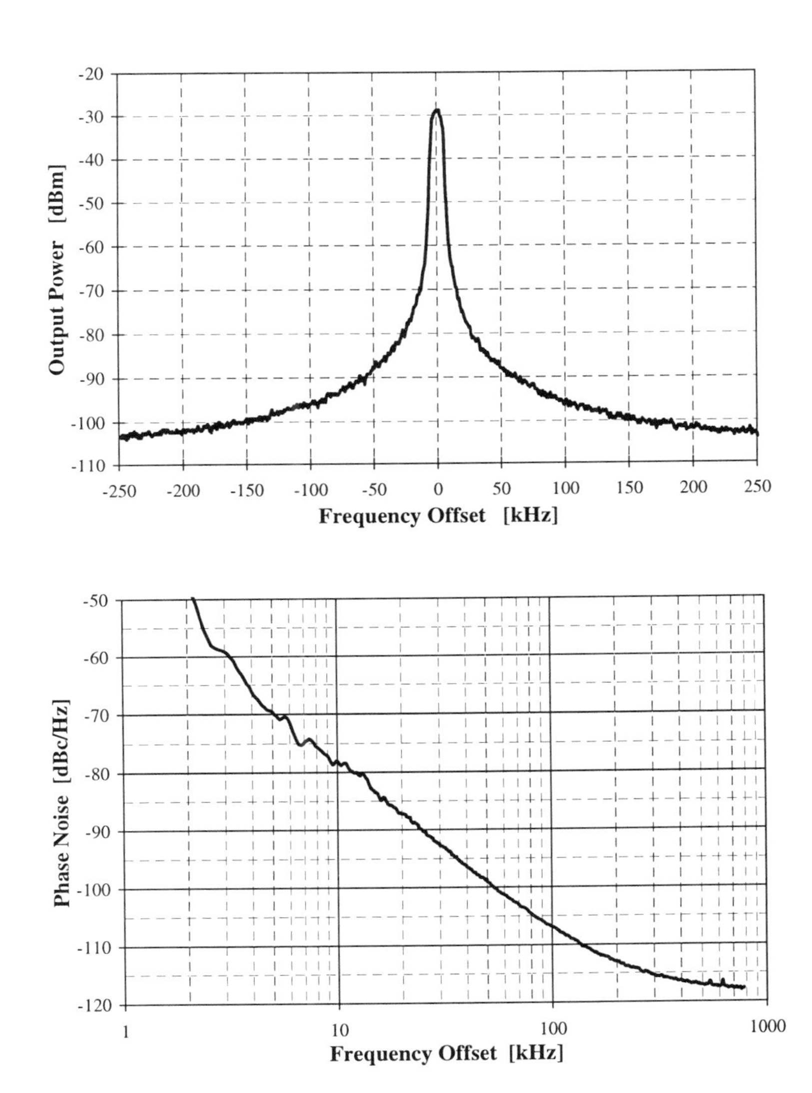

Figure 5.22. Measured planar-LC DCS-1800 VCO phase noise : (a) Output spectrum for a carrier frequency of $1.99\ GHz$ and resolution bandwidth of $10\ kHz$; (b) Logarithmic plot

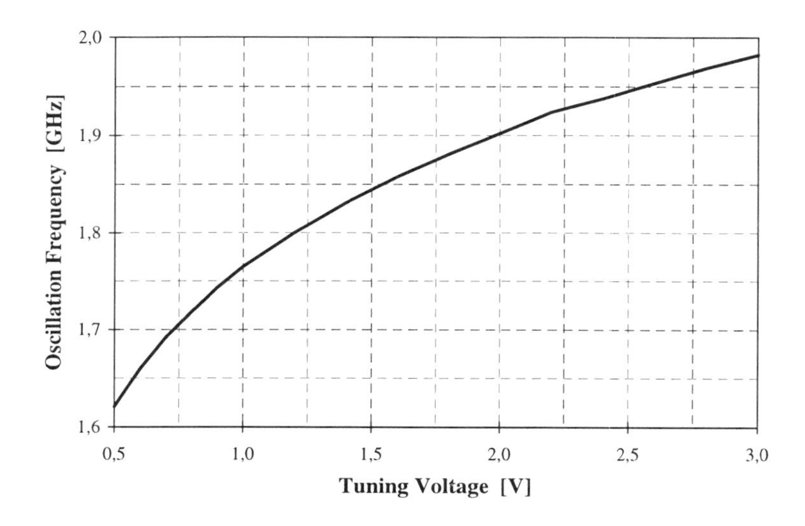

Figure 5.23. Measured planar-LC DCS-1800 VCO tuning characteristic

guideline is only valid for conductive substrates, as the substrate currents in lowly-doped wafers are already small.

Based on these guidelines and the finite-element simulations, two standard CMOS designs have been made. The first one is a worst-case design, in the sense that the technology used has only two metal layers available and uses heavily doped wafers. Nevertheless, the inductor coil has been sized such that an optimum has been reached with respect to metal losses, silicon losses, inductance value and phase noise. The resulting VCO has a 1.8-GHz center frequency, consumes only 6 mW, and has a phase noise of -116 dBc/Hz at an offset of 600 kHz. As can be seen from the data in table 3.2, this is by more than 10 dB the lowest phase noise obtained for a planar-LC oscillator on a conductive substrate.

A second design has been done in a 0.4-μm CMOS technology that doesn't use epi-wafers. The absence of substrate losses has enabled to go for the real thing : a set of two integrated VCOs with planar inductors that achieve the GSM and the DCS-1800 phase noise specs. A completely symmetrical octagonal coil shape has been used in both designs. The resulting phase noise is as low as -108 dBc/Hz at 100 kHz offset for the GSM design and -113 dBc/Hz at 200 kHz offset for the DCS-1800 design. This last number is only 2 dB worse than the bonding wire VCO result. The power dissipation is 9 and 11 mW, respectively, and the full system frequency band is covered by their tuning range.

Although this last design benefits from the small transistor features available, a very important problem of future submicron technologies has come up, i.e. the $1/f$ noise. When the gate length becomes smaller and smaller, the $1/f$ noise increases because of the smaller transistor area. This is enhanced by the technology's Spice pink noise factor KF that is generally larger for processes with thinner oxides, as e.g. the surface scattering responsible for $1/f$ noise increases. In our design this was noticed at low VCO control voltages, when the nonlinearities in the tuning junction capacitor are larger and the upconversion process of low-frequency $1/f$ noise has become more dominant.

These two examples have demonstrated that by proper analysis of integrated planar inductors, it is possible to design VCOs that fulfill the phase noise requirements of modern mobile communication systems. Previously, oscillators were designed using inductors that were characterized in advance and that were also used in other applications. Not surprisingly, the results were poor. We have demonstrated that by thorough finite-element analysis, is it possible to find the optimal inductor coil that will do the job.

6 HIGH-FREQUENCY CMOS PRESCALERS

6.1 INTRODUCTION

As already stated in the introduction, two high-frequency building blocks of a PLL frequency synthesizer require extra effort to be implemented in a standard CMOS process. The voltage-controlled oscillator has already been discussed in chapters 3 to 5. This chapter deals with the other circuit that also operates at the full output frequency, i.e. the frequency divider. The several types of frequency dividers have already been listed is section 2.4.4. A divider that can handle all integer division numbers must be a programmable counter. But because of the large amount of logic required to make a function like that, its speed is limited, and operation at more than 1 GHz in a standard CMOS process is certainly impossible. Therefore a prescaler can be positioned before the normal frequency divider to reduce the required speed. This prescaler circuit divides by a fixed number, usually a power of 2, and can operate at higher frequencies since the critical path can be kept short. The required speed of the subsequent stages is lowered by the prescaler division number, but also the smallest frequency step has increased by that number. For a constant frequency resolution, the reference frequency must be lowered, which of course has its disadvantages. A dual-modulus prescaler can divide by *two* different numbers, and is used together with a pulse- and a swallow-counter (see figure 2.25) to resolve this problem. Also in the design of fractional-N division systems, dual-modulus prescalers are used.

In this chapter we will investigate the implementation aspects of such a Dual-Modulus Prescaler (DMP) in a standard CMOS process. The main issue is of course to achieve a speed as high as possible, at a reasonable power consumption. Two innovations have been used to realize his goal. First, a new DMP architecture is presented in section 6.2. It is based on a switching strategy between the 4 outputs of a differential master/slave flipflop, which are spaced 90° apart. Therefore we call it the *phase-switching prescaler* architecture. Its main advantage is that it requires only an asynchronous divide-by-4 prescaler in its input stage. This can be operated at higher speeds than a synchronous divide-by-4/5 circuit that is used in a conventional topology. So this new architecture allows dual-modulus operation at the same speed as an asynchronous divider. Secondly, a high-speed divide-by-2 toggle flipflop is optimized for this application. This is demonstrated in section 6.3 with the design of a dual-modulus divide-by-128/129 prescaler in a 0.7-μm CMOS process. A redesign of that circuit in a 0.4-μm technology is presented in section 6.4. An improvement has been made to create an eight-modulus prescaler that can handle all integer division numbers from 64 to 71. In the aimed application, i.e. a DCS-1800 frequency synthesizer, this prescaler can function as the only block in the frequency divider. So no configuration with a pulse- and a swallow-counter is necessary, as the full frequency band of interest is covered with these division numbers. Finally, section 6.5 draws some conclusions from this chapter.

6.2 PHASE-SWITCHING DUAL-MODULUS PRESCALER ARCHITECTURE

This section presents the block diagram of the conventional implementation of a DMP, as well as the operating principle of the new high-speed DMP topology. Some interesting extensions of the basic dual-modulus operation are also discussed shortly.

6.2.1 *Conventional Topology*

A conventional high-speed dual-modulus divide-by-128/129 prescaler generally consists of a synchronous divide-by-4/5 part and an asynchronous divide-by-32 part as shown in figure 6.1 [Rogen CICC94]. The synchronous divider is the only part of the circuit operating at the maximum input frequency, and was already depicted in figure 2.24. Most of the time, its control signal $Ctrl$ is low, so the output frequency $F4$ is determined by the loop over the first two D-flipflops and equals $1/4^{th}$ of the input frequency. This frequency is divided by 32 in the asynchronous divider to obtain an output frequency F_{out} equal to $F_{in}/128$. The divide-by-129 operation is enabled by setting the $Mode$-input high. When the outputs of all flipflops of the asynchronous

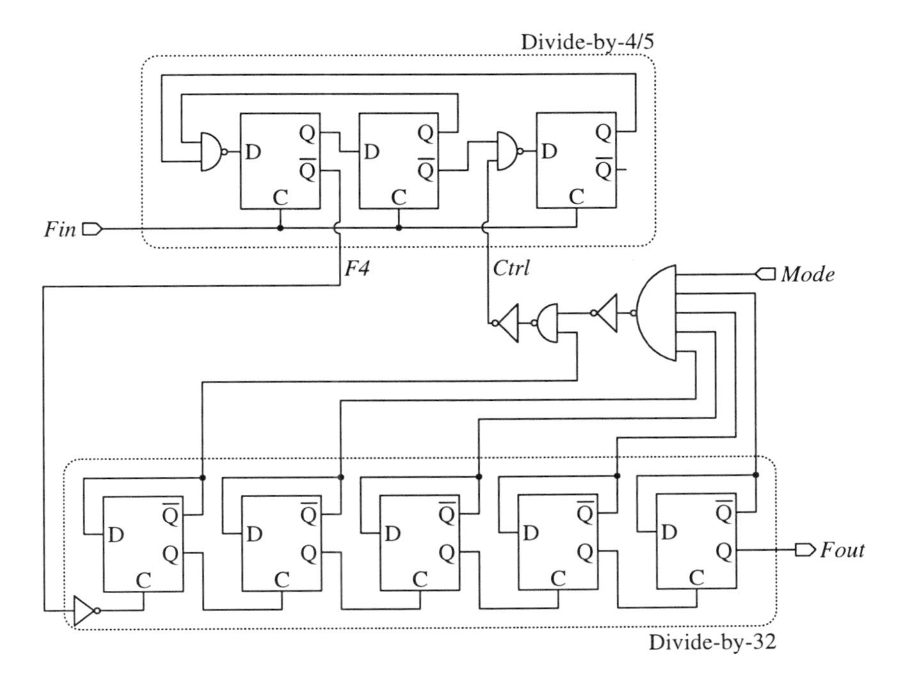

Figure 6.1. Conventional dual-modulus prescaler architecture

divider are high, i.e. once every period of the output signal F_{out}, the $Ctrl$-signal becomes high. This causes the loop in the synchronous divider to be momentarily closed over three flipflops instead of two. This extra delay is equivalent to a divide-by-5 operation. So the prescaler divides once by 5 and 31 times by 4, which results in a division by 129.

The problems with this prescaler topology are of course situated in the synchronous divider. It is the only part of the circuit operating at the maximum frequency, the rest of the circuit runs at maximum $1/4^{th}$ of the input frequency. The synchronous divider contains 3 high-frequency fully functional D-flipflops. These flipflops are the reason for a lot of power consumption and clock load. Moreover, the NAND-gates in the critical path of the synchronous divider loop will cause a decrease in the maximum input frequency. Clever design can reduce this effect by embedding the NAND-gate into the first stage of the flipflop, but can never eliminate it completely. Therefore, a dual-modulus prescaler with this architecture will always have a smaller operating speed than an asynchronous frequency divider, e.g. 1.4 GHz with respect to 1.57 GHz [Rogen CICC94]. Moreover, circuits can be optimized for a divide-by-2 operation such

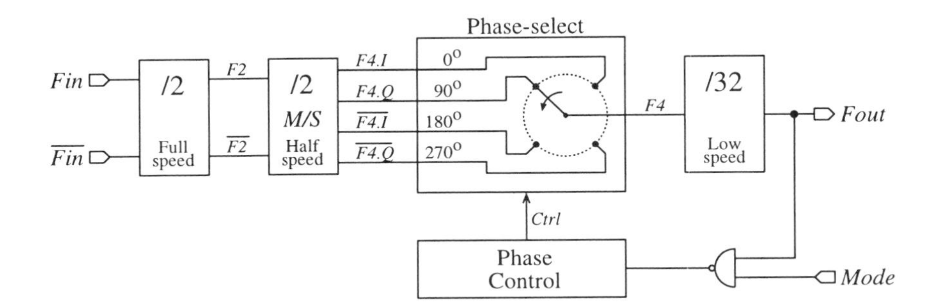

Figure 6.2. Phase-switching dual-modulus prescaler architecture

that their speed can be roughly a factor of two higher than synchronous dual-modulus dividers [Razav CICC97].

6.2.2 Phase-Switching Topology

The new dual-modulus prescaler topology proposed here solves all these problems. The block diagram of a basic configuration is shown in figure 6.2. It consists of a chain of 7 pure divide-by-2 circuits interrupted by a phase-select block. Since the frequency-limiting first stage is only one toggle-flipflop, input frequencies as high as asynchronous dividers can be obtained. The dual-modulus operation is based on the 90-degrees phase relationship between the outputs of the master and the slave of a Master/Slave (M/S) D-flipflop.

The phase-switching prescaler operates as follows. The differential input signal F_{in} is fed to a first high-speed divide-by-2 flipflop. This flipflop is the only one operating at the full input frequency. It must not be a fully functional D-flipflop, but can be optimized for divide-by-2 operation. The resulting signal $F2$ is once again divided by a second high-speed divide-by-2 flipflop. For the dual-modulus operation, this flipflop must be of the master/slave kind as shown in figure 6.3(a). Four output signals result : the differential output of the master ($F4.I$ and $\overline{F4.I}$) and the differential output of the slave ($F4.Q$ and $\overline{F4.Q}$). When the $Mode$-input is low, the phase-control block is disabled and its output signal $Ctrl$ will be constant. Therefore, the phase-select circuitry simply picks one of its four input signals and connects it to the input of the asynchronous divide-by-32 block. This block is a chain of 5 divide-by-2 flipflops operating at a relatively low speed (maximum $1/4^{th}$ of the input frequency). The resulting output frequency is thus a factor of $2 \times 2 \times 32 = 128$ smaller than the input frequency.

The divide-by-129 operation is enabled by setting the $Mode$-input high. The phase-control block is now working. On every positive edge of the output signal F_{out}, the

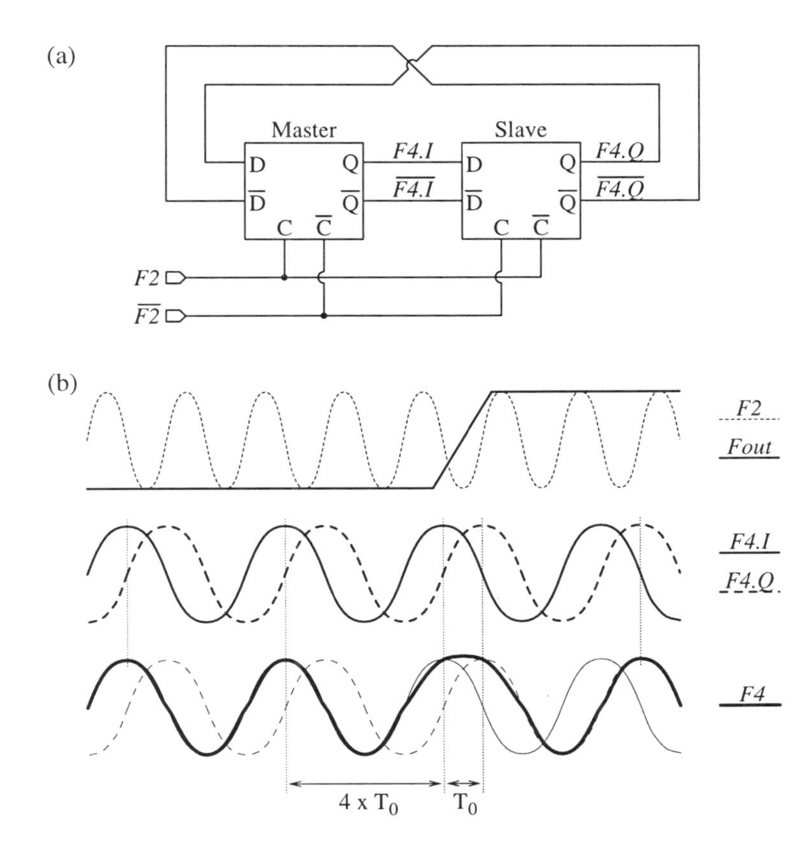

Figure 6.3. Phase-selection principle in the phase-switching DMP architecture : (a) Master/Slave divide-by-2 flipflop; (b) waveforms for a switch causing a 90-degree delay

control signal $Ctrl$ will be changed in such a way that the phase-select block will connect $F4$ to that signal of the four flipflop outputs that is 90° *delayed* with respect to the present signal. So e.g. when $F4$ is initially connected to $F4.I$, after the rising edge of F_{out} a connection will be made to $F4.Q$. This is shown in figure 6.3(b). Since the signal $F4.Q$ lags the signal $F4.I$ by 90°, the signal $F4$ will be delayed due to this operation. The delay is equal to 90° of a signal with period $4 \times T_0$, or equal to T_0. So also the output period is increased with this delay and now equals $128 \times T_0 + T_0$. The prescaler division factor is now 129, and a dual-modulus operation is obtained by using only divide-by-2 toggle-flipflops.

This new phase-switching prescaler architecture is thus inherently faster than a conventional one. Instead of 3 full-speed D-flipflops that must be able to perform all logical functions, only one full-speed divide-by-2 flipflop is needed. No extra logic is needed as in a 4/5-divider. These are basically the two aspects that allow to achieve very high input frequencies with this prescaler. First, the extra delay that is added to the critical path by the dual-modulus logic of a divide-by-4/5 circuit is not present here. This will certainly result in smaller delay times and higher speed. Secondly, the full-speed flipflop is not a fully functional D-flipflop, but only performs a divide-by-2 function. This will allow the use of special transistor schematics for this block, as one knows that the data input signal is 01010101 .... There are never two consecutive ones or zeros, so the flipflop must not be able to handle this input. This will become more clear in the circuit design example of section 6.3.

6.2.3 Variations on the Basic Topology

A lot of variations on this basic architecture can be developed, in order to create prescalers that can handle division factors other than N and $N + 1$. A first option that comes to mind is to make a prescaler that divides by N and $N + 1/2$. How to do this is straightforward, just omit the first divide-by-2 block and put the phase-selection such that it operates on the signals at half the input frequency instead of one-fourth of F_{in}. Being able to divide by not only integer numbers but also half integers can be a big advantage in designing PLL synthesizers because it enables to increase the reference frequency without reducing the frequency resolution. But of course the circuit design will become more difficult, as the phase-selecting circuitry must now operate at twice the previous speed. In the two designs presented later in this chapter, this was not feasible. Another way to make an $N + 1/2$-divider is to use eight signals at $F_{in}/4$ coming from two half-speed M/S divide-by-2 flipflops driven by a M/S full-speed divider. This is shown in figure 6.4. If the first flipflop also has a master and a slave section, one can use the four quadrature signals at half the frequency to drive two separate half-speed dividers. One is driven by the in-phase signals $F2.I$ and $\overline{F2.I}$, and produces four output signals $F4i.X$ as indicated in the figure. The other one is driven by the quadrature signals $F2.Q$ and $\overline{F2.Q}$ and produces four output signals $F4q.X$. As the in-phase and quadrature signals $F2$ are spaced 90° apart, the outputs of both blocks $F4i.X$ and $F4q.X$ are spaced 45° apart, which results in eight signals at one-fourth of the input frequency, equally divided over the 360-degrees period. By creating a phase-selecting circuit that introduces a 45-degree delay for each period of the output signal, this output signal will be delayed by half an input period. So an extra division factor of 1/2 is realized. Of course, since now two half-speed dividers and a larger phase-selection circuit are needed the power consumption due to this extra feature will increase. And a major problem that is already indicated in figure 6.4 is the fact that without any precaution, one is not sure about the relative phase of the two

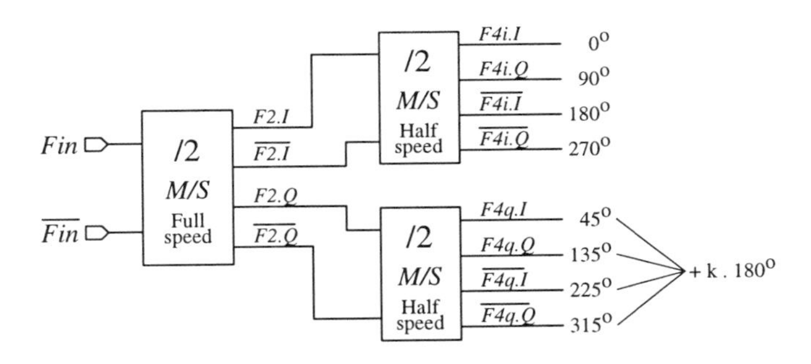

Figure 6.4. Phase-switching prescaler for divide-by-N+1/2 operation

half-speed dividers with respect to one another. The initial states of the flipflops are not known, and hence their output can be 1010... or 0101..., which gives of course a 180-degrees phase shift in the output signals. So both dividers must be coupled in some way in order to control their relative phase, or a startup-circuit must be added in order to assure the initial states of the flipflops.

The second variation on the basic phase-switching prescaler topology of figure 6.2 is the realization of more than two division moduli. In fact, if we switch the phases more than once per period of the output signal, we will divide by $N + s$, with s the number of phase switches. A suitable block diagram is shown in figure 6.5. The simple NAND gate that triggered the phase-control block in the basic configuration is now replaced by a larger modulus-control block. This block uses the 5 intermediate outputs of the asynchronous divide-by-32 circuit, and is controlled by a $Mode$ signal of $n_s = \log_2 s$ bits. Based on these signals, it can deliver a number of pulses on its output $Next$. The number of pulses is determined by the $Mode$-word and their positioning is aligned to the intermediate signals $/8, /16, \ldots, /128$. The phase-control block is designed such that it switches the phase of $F4$ for every pulse it receives on its input. An implementation will be presented in section 6.4.

The limitations encountered in the design of these new topologies are mainly speed limitations. One has already been mentioned, i.e. to make a division by $N + 1/2$ it would be preferable to run the phase-selection at half the input frequency. As this is not always feasible in very-high speed applications, we have to fall back on e.g. the solution depicted in figure 6.4. Another limitation also originates from the phase-selection block and has an influence on the smallest division factor realizable. It is important that the prescaler division factor is not too high, because the minimum division factor in a frequency divider with DMP and P- and S-counters (see figure 2.25) is proportional to the square of this number. Theoretically, we can easily make a divide-by-4/5 phase-switching prescaler by omitting the low-speed asynchronous divide-by-32 block and controlling the NAND gate with the $F4$ signal. However, in a 4/5 DMP the phase

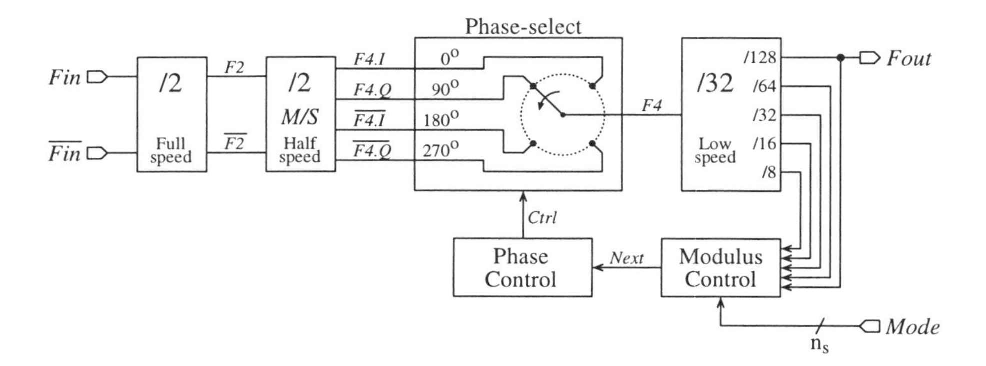

Figure 6.5. Phase-switching multi-modulus prescaler

switches happen with very short intervals and this can sometimes cause problems in the phase-selector, as the transients that appear in its circuits while switching have not settled yet before the next switch must happen. This will be illustrated in the design example of the next section.

6.3 A DUAL-MODULUS DIVIDE-BY-128/129 PRESCALER IN 0.7-μm CMOS

In the same 0.7-μm CMOS technology as the bonding wire VCO and the first planar-inductor VCO, an implementation of the new prescaler architecture has been realized. This section presents the circuit diagrams of the several building blocks, stresses some problems encountered in the implementation, and reports the measurement results of the realized circuit. To be compatible with the trend to lower the supply voltage in low-power applications and submicron processes, a 3-V supply was specified. The block diagram of this prescaler is the basic topology shown in figure 6.2.

6.3.1 Circuit Design

6.3.1.1 Full-Frequency Divide-by-2. The frequency-limiting building block in the architecture is of course the first divide-by-2 block. A D-flipflop with its negative output ($\overline{Q}$) connected to its input (D) performs this function. Several high-performant CMOS D-flipflops are published in open literature. Some of them will be briefly discussed here.

A first example is a series of designs based on the dynamic True Single Phase Clock (TSPC) circuit technique [Ji-Ren JSSC89]. A slightly modified and optimized D-flipflop is presented in [Rogen CICC93, Rogen CICC94, Huang JSSC96]. To make an

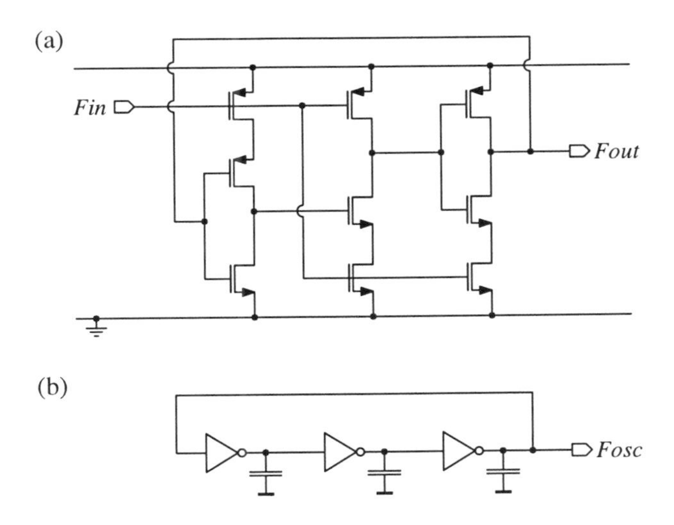

Figure 6.6. Analogy between (a) Dynamic TSPC CMOS toggle-flipflop and (b) Three-inverter ring oscillator

estimation of the maximum frequency obtainable with the dynamic flipflop of [Rogen CICC93], we can analyze the oscillation frequency of a three-inverter ring oscillator. This analogy is shown in figure 6.6. The toggle-flipflop can be regarded as being a three-inverter ring oscillator with some additional control transistors that will regulate the oscillation frequency to a certain value, i.e. half of the input toggle frequency. These control transistors will slow down the circuit, so the maximum input frequency obtainable with the dynamic D-flipflop is less then twice the oscillation frequency of the ring oscillator.

In our 0.7-μm CMOS technology, the oscillation frequency of the ring oscillator is only 1.5 GHz at a 5-V power supply, and only 0.8 GHz at 3 V. This means the maximum input frequency of the toggle-flipflop will be lower than twice this value, or 1.6 GHz at 3 V. To compare this with the results obtained in [Rogen CICC93], the simulation is also performed for 1.2-μm design rules. For a 5-V supply, this renders an oscillation frequency of 0.8 GHz. So the possible speed enhancement by going from a 1.2-μm process to 0.7 μm is completely eliminated by lowering the supply voltage from 5 V to 3 V.

Another example are the level-triggered latches of [Forou JSSC95]. They are also not suited for our purposes as they operate with a 5-V power supply and their speed drops rapidly at lower voltages. And in their maximum-speed bias conditions, the input frequency range becomes very small.

A very-high speed divide-by-2 implementation reports a maximum input frequency of 13.4 GHz [Razav ISSCC94a, Razav JSSC95], but it uses a partially scaled 0.1-μm CMOS technology. The circuit also uses a master/slave configuration, and the flipflop circuit schematic is a first example of how the D-latch can be optimized for use this divide-by-2 operation instead of a general digital logic circuit. The D-latch does not disable its input devices in the hold-mode. This would otherwise certainly pose timing problems, but it does not prevent the divider from functioning properly in this configuration.

Other prescaler designs use special technologies to achieve the required high speeds, e.g. SIMOX/SOI (Separation by IMplanted OXygen / Silicon On Insulator) substrates [Fujis JSSC93, Kado JSSC93]. This option is not considered in this work.

The circuit used in this design tries to achieve a speed enhancement by limiting the output swing of the flipflop, and thus reducing the time required to switch the output signal from a low level to a high level. The circuit is based on a standard Master/Slave ECL D-flipflop. A straightforward CMOS implementation of one section is shown in figure 6.7(a). All bipolar npn transistors have been replaced by NMOS transistors, and the resistor normally used as a load for the circuit is now implemented as a diode-connected PMOS transistor. To increase the maximum toggle frequency, the current source I_{bias} can be omitted, and the source of transistors $M1$ and $M2$ is connected directly to the ground rail. Omitting the current source implies that, in order to drive the input transistors sufficiently into and out of saturation, the input signal swing must be large enough. In this design, a 1.5-V_{ptp} input amplitude was used. Since this prescaler can be driven directly by the VCO, this poses no problem because the VCO output swing is already maximized for low phase noise. So the circuit exploits the speed enhancement possible by the reduction in voltage swing from input to output. Also the necessity of differential input signals poses no extra cost, since the VCO will always be implemented differentially. This circuit was already implemented successfully in a 4-V fixed-modulus prescaler [Crani JSSC95]. Simulations indicate a speed improvement of 20% over a standard implementation. Since the second divide-by-2 operates at only half the frequency, the smaller output swing poses no problems.

To fit this circuit into a 3-V power supply, another measure must be taken. To avoid the V_{GS} voltage drop across the diode load transistors $M7$ and $M8$, they must be folded to ground. They can than be replaced with NMOS transistors, which is another small advantage because of the smaller parasitic capacitances. This final circuit is shown in figure 6.7(b).

The sizing of this circuit is not straightforward. Because the circuit is biased with the DC level of the VCO signal, and not with the normal current source, care has to be taken to ensure proper operation over all process variations. This was investigated with numerous Monte-Carlo simulations. Therefore, a Spice model library was developed that contained the dependencies of the model parameters on the possible process variations. This library was used the generate statistically random models, which were

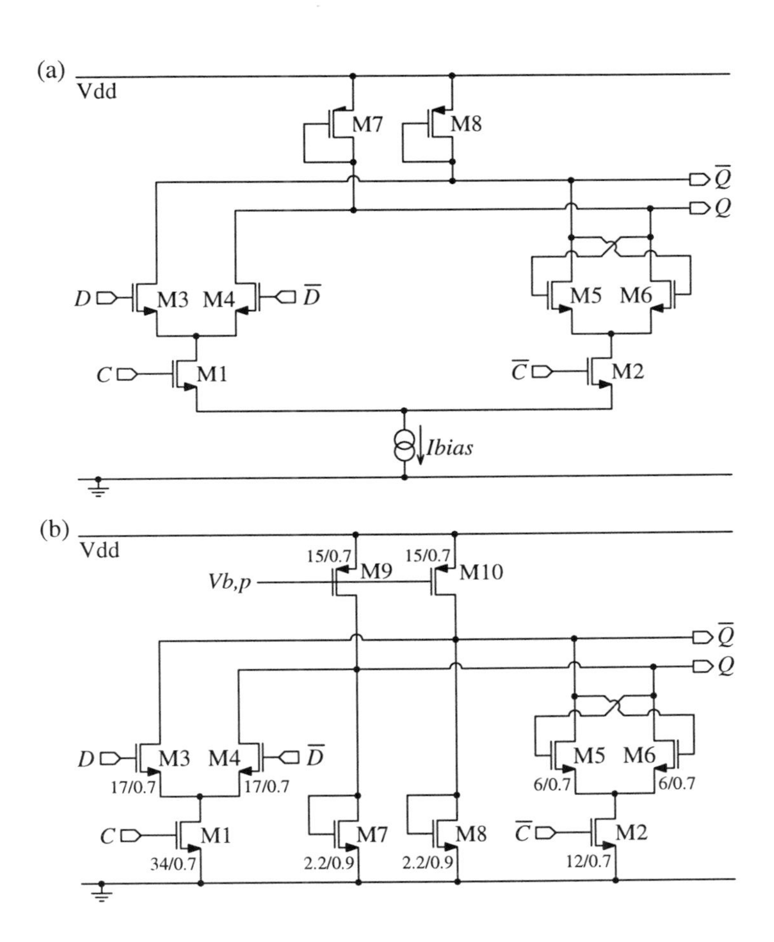

Figure 6.7. (a) CMOS implementation of an ECL flipflop; (b) One section of the high-frequency D-flipflop in $0.7\text{-}\mu m$ technology

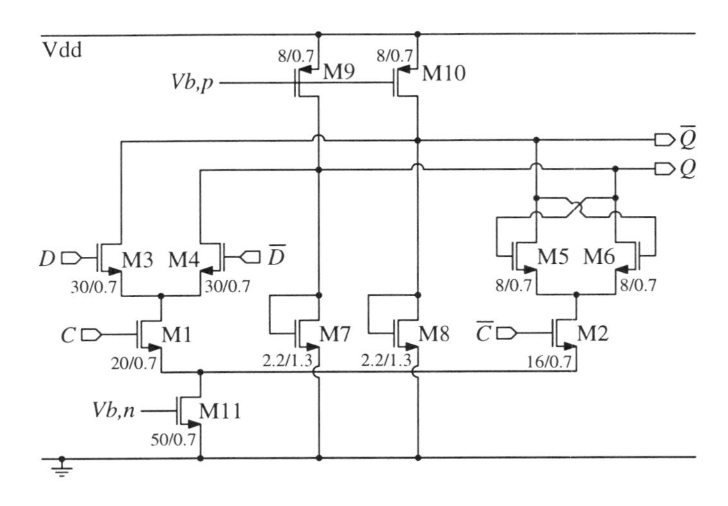

Figure 6.8. Half-frequency divide-by-2 flipflop in $0.7\text{-}\mu m$ technology

used to simulate the circuit under varying process conditions. The 3-σ variation of e.g. the threshold voltage V_T was 150 mV. The model parameters were generated consistent with reality, e.g the oxide thickness of the NMOS and the PMOS model was kept the same. The operating temperature was also varied.

So the circuit was not sized to obtain a maximum operation speed with typical model parameters, but to be robust for process variations and temperature changes. The operating point was chosen to guarantee a reasonable yield (e.g. 95%) at a somewhat reduced frequency. Of course, the yield under real processing conditions can not be predicted safely with this procedure, but certainly an improvement over a design with only typical parameter simulations is achieved. The resulting transistor sizes are indicated in the figure. The optimized circuit has a simulated maximum toggle frequency of 2 GHz at 3 V. The output amplitude is approximately 0.7 V_{ptp}.

6.3.1.2 Half-Frequency Divide-by-2. The second divide-by-2 circuit in this prescaler architecture must be a M/S toggle flipflop, as we need the four differential quadrature outputs at one-fourth of the input frequency. Its design is based on the first divider, but the bias current source cannot be omitted. This is because we have to cope with the smaller input amplitude and the higher DC level of its input signal, which is the output signal of the first divider-by-2. The circuit schematic with the transistor sizes is shown in figure 6.8. Since this divider operates at half the input frequency, the speed enhancement that resulted from omitting the bias current source is no longer necessary. The output amplitude is approximately 0.5 V_{ptp}.

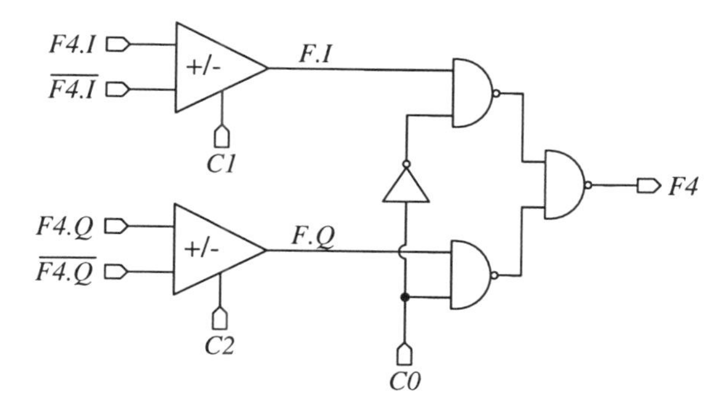

Figure 6.9. General circuit diagram for phase-selection

6.3.1.3 Phase-Selection. The phase-select circuitry is shown in figure 6.9. The selection of the correct signal is done in two stages. In a first stage both the in-phase (I) and quadrature (Q) signals are amplified in a differential-to-single-ended amplifier. This is necessary because the amplitude of the output signals of the second divide-by-2 circuit is only 0.5 V_{ptp}. The amplifier can be switched between positive and negative amplification, thereby making a selection between the positive signals ($F4.I$ and $F4.Q$) or the negative signals ($\overline{F4.I}$ and $\overline{F4.Q}$). The control signals $C1$ and $C2$ are used for this.

The circuit schematic of this amplifier with transistor sizes is shown in figure 6.10. By changing the polarity of the control signals C and $\overline{C}$, the bottom current mirror can be coupled to the rest of the circuit in a positive or a negative mode. Because of the high DC level of the input signals, this configuration was not possible with a PMOS current mirror directly above the input differential pair. The currents had to be mirrored down first.

It is in this amplifier that the previously mentioned limit on the minimum realizable division factor is situated. The operating speed ($F_{in}/4$) is at the edge of what is capable in this technology, and therefore its output waveform is not clipped between the power supply and ground. In fact, this amplifier has trouble generating a large-amplitude output at this frequency. Once settled, the output is large enough to drive the following stages of the phase-selector, but at the switching instances of its control signal $C1$ or $C2$, transients appear that temporarily corrupt its correct behavior. It takes a while before the DC-level of the amplifier output is settled to the correct voltage, i.e. the threshold of the NAND gate following this block. Since this switching takes place when the phase-control block selects the output of the other amplifier, it does not affect the correct operation of the prescaler. But when the phase switches follow each other rapidly, it might cause problems when the settling isn't finished yet at the time when the amplifier is again selected.

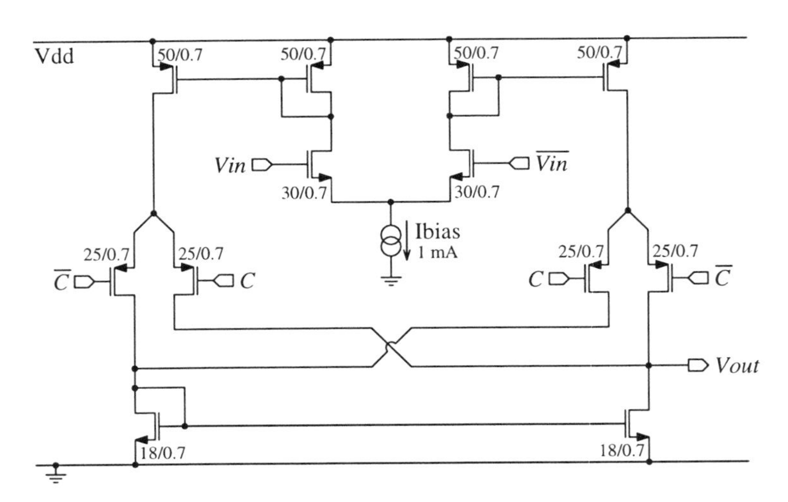

Figure 6.10. Switchable amplifier for phase-selection

A selection between the two remaining signals $F.I$ and $F.Q$ is made with three simple NAND-gates and the control signal $C0$ (see figure 6.9). A very important aspect of this circuit, is what happens at the transfer from one selection to another. Uncareful design can cause spikes in the signal $F4$, resulting in an improper division by the divide-by-32 block. A smooth conversion in $F4$ must be guaranteed for all possible variations in processing or temperature and for all input frequencies. This is only important for the NAND-gates controlled by $C0$, because the control signals $C1$ and $C2$ are changed when the second stage of the circuit has selected the other signal (i.e. when $C0$ selects $F.I$ to be connected to $F4$, $C2$ is changed). The transients that occur when switching the amplifiers from one amplification mode to another can therefore never cause spikes in the output.

For high input frequencies, the risk of creating spikes doesn't exist, because the NAND-gates don't react fast enough. However, for lower input frequencies, there is a possibility of spikes. A simulation of this is shown in figure 6.11. The following waveforms are shown from top to bottom : the control signal $C0$, the signals $F.I$ and $F.Q$ (notice the 90° delay of $F.Q$), and the resulting signal $F4$. The simulation is performed with fast transistor models and at an operating frequency of 250 MHz, which corresponds to an input frequency of 1 GHz. When $C0$ is low, the high-to-low transitions in $F4$ are determined by the signal $F.I$. After a $0 \rightarrow 1$ change of $C0$, the transitions are determined by $F.Q$. This is indicated on the figure. However, if the control signal $C0$ has a very steep slope, $F.I$ can be deselected (and the signal $F4$

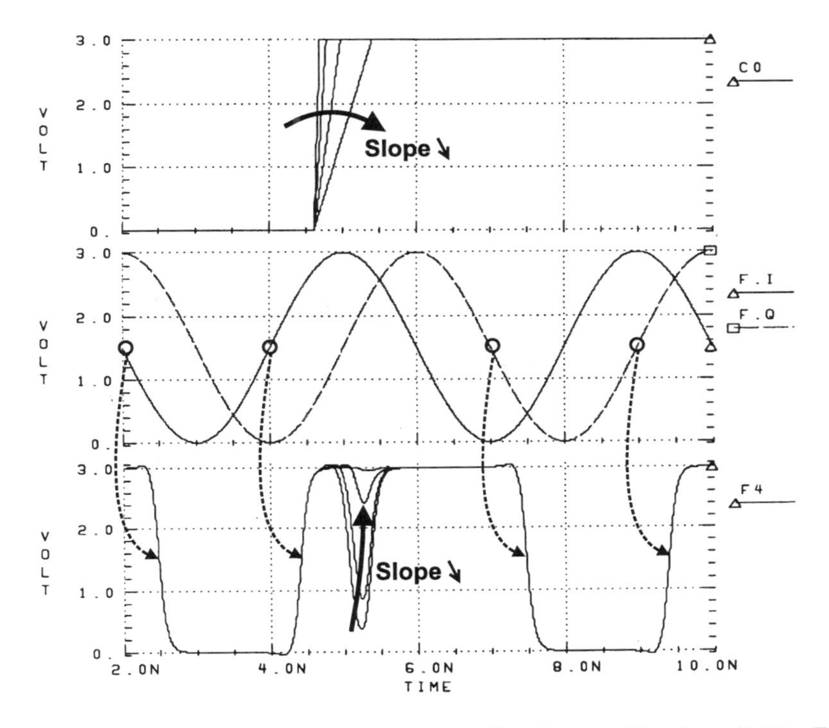

Figure 6.11. Illustration of the risk of spikes in $F4$ at the transition from $F.I$ to $F.Q$

goes from high to low) before $F.Q$ is high enough. This will cause a negative spike in $F4$, as shown on the bottom curve of figure 6.11.

The presence of this effect is dependent on the exact time of arrival and on the slope of the change in $C0$. The spike only occurs for a very critical time of arrival. But since the delay in $C0$ cannot be controlled nor guaranteed, the possibility of having exactly that arrival time cannot be ruled out. One could think of controlling the control signal arrival time by inserting a clocked D-flipflop into the path of $C0$. This flipflop could be clocked by e.g. the signal $F4$. But even in this case the delay of the flipflop is not guaranteed, and the risk of spikes is not eliminated.

However, the problem is solved more easily by lowering the slope of $C0$. In figure 6.11 it can be seen that the spike disappears as the slope of $C0$ lowers. So a very small buffer inverter that has a large rise- and fall time is used to drive the control signals, in order to limit the slope. As for process variations, the range over which the slope can vary without the risk of spikes is very large. So the buffer inverter is designed such that a smooth transition in $F4$ for all possible arrival times of $C0$ and for all possible transistor processing variations and operating temperatures.

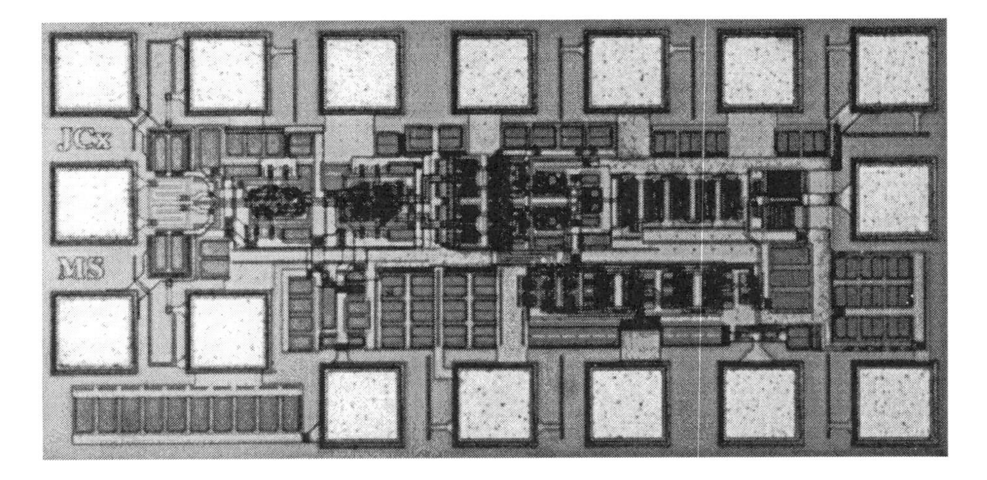

Figure 6.12. Microphotograph of the 0.7-μm CMOS divide-by-128/129 DMP IC

6.3.1.4 Low-Frequency Divide-by-32. This divider is an asynchronous divider that consists of a chain of five toggle-flipflops. The maximum operating frequency is $1/4^{th}$ of the input frequency, so the dynamic flipflop proposed in [Rogen CICC93], as already shown in figure 6.6(a), can be used.

6.3.2 Measurement Results

The circuit has been implemented in a standard 0.7-μm CMOS process. A chip microphotograph is shown in figure 6.12. The die size, including bonding pads, is $1150 \times 550\,\mu m^2$. Measurements were made at room temperature with the chip bonded on a ceramic substrate. The maximum input frequency of this technique is limited by the parasitic bonding wire inductance and bonding pad capacitance. The input signals were terminated with on-chip 200-Ω resistances. This results in a small overshoot in the input transfer characteristic in comparison with a 50-Ω termination, but the maximum input frequency is extended. The required differential input signal was easily made by inserting a small delay into one of the two inputs. This delay is not critical, since the applied phase shift (normally 180° for a perfect differential signal) can be varied by more than 50°.

The IC consumes, output buffer not included, 8 mA. This is divided over the several building blocks as follows : full-speed divide-by-2 : 2.5 mA; half-speed divide-by-2 : 1.5 mA; phase-selection : 2.5 mA; low-speed divide-by-32 : 1.5 mA.

The measured maximum input frequency is up to 1.75 GHz at 3 V. The output signal for both division factors, together with the 1.75-GHz input signal, is shown

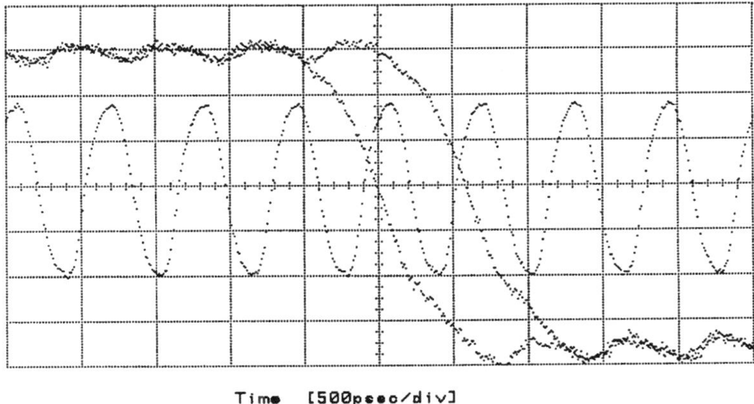

Figure 6.13. Input and output waveforms for both division factors at 1.75 GHz

in figure 6.13. It can clearly be seen that the divide-by-129 signal has a high-to-low transition that is delayed by one period of the input signal with respect to the divide-by-128 signal. This measured result is about 10% less than what could be expected from simulations. This difference is probably due to insufficient modeling of the high-frequency effects of the transistors.

As for the influence of processing variations, these could not be measured completely since only a limited number of samples were available. However, an indication of the robustness of the circuit is given by the influence of e.g. the power supply voltage or the externally supplied biasing current on the operating frequency. The maximum input frequency vs. supply voltage is shown in figure 6.14. Although this circuit was optimized for 3-V operation, an input frequency as high as 2.65 GHz was measured with a 5-V power supply.

The phase noise of this prescaler was measured with the HP3048A phase noise measurement system, with a configuration as shown in figure 6.15(a). The measurement results for a 1.28-GHz input signal are shown in figure 6.15(b). This frequency was chosen because a very stable 10-MHz reference was available. The measured output phase noise is -111 dBc/Hz at 100 Hz offset, -131 dBc/Hz at 1 kHz offset, and flattens for higher offset frequencies to a noise floor of -142 dBc/Hz.

6.4 AN EIGHT-MODULUS PRESCALER IN 0.4-μm CMOS

This second design example discusses a slightly modified version of the previous one. It has been implemented in a deeper submicron technology, which has enabled to cut down the power consumption, and the number of possible division factors is increased

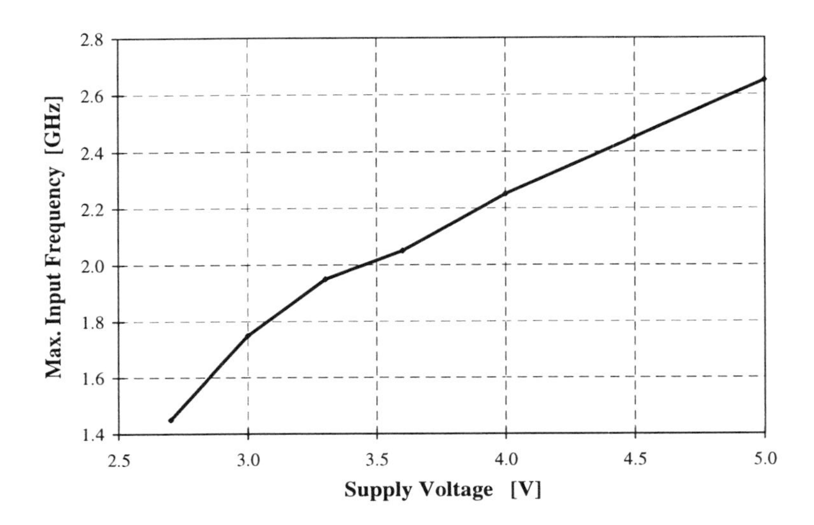

Figure 6.14. Measured maximum input frequency vs. supply voltage

to eight as already explained in figure 6.5. The reason for doing this is the fact that now the complete frequency band of the DCS-1800 system (i.e. $1.71 - 1.88$ GHz) can be covered with a PLL with a frequency divider consisting of only this prescaler. Starting from a 26.6-MHz reference, multiplication by 64 yields an output frequency of 1702.4 MHz, which is lower than the lowest frequency required. And multiplication by 71 results in 1888.6 MHz, which is higher than the highest frequency. Of course, to achieve the required frequency resolution of 200 kHz, addition of the fractional-N technique as discussed in section 2.4.4.3 is necessary.

6.4.1 Circuit Design

This first section describes some of the circuit implementations of the building blocks of the improved design. Because of the smaller lithography used, faster transistors can be realized, even at low $V_{GS} - V_T$ values. This fact and the smaller threshold voltage values have allowed to change some of the schematics to a more optimal configuration and to cut the power consumption.

In the full-speed divide-by-2, it was no longer necessary to fold the load transistors $M7$ and $M8$ to ground. The PMOS threshold voltage is low enough (0.7 V, instead of 0.95 V in the 0.7-μm process), and because of the small channel length the use of a PMOS transistor instead of an NMOS poses no speed limit. The circuit schematic, with the transistor sizes used, is shown in figure 6.16. Again, the transistor sizes are determined in order to ensure a sufficient yield over all possible process variations.

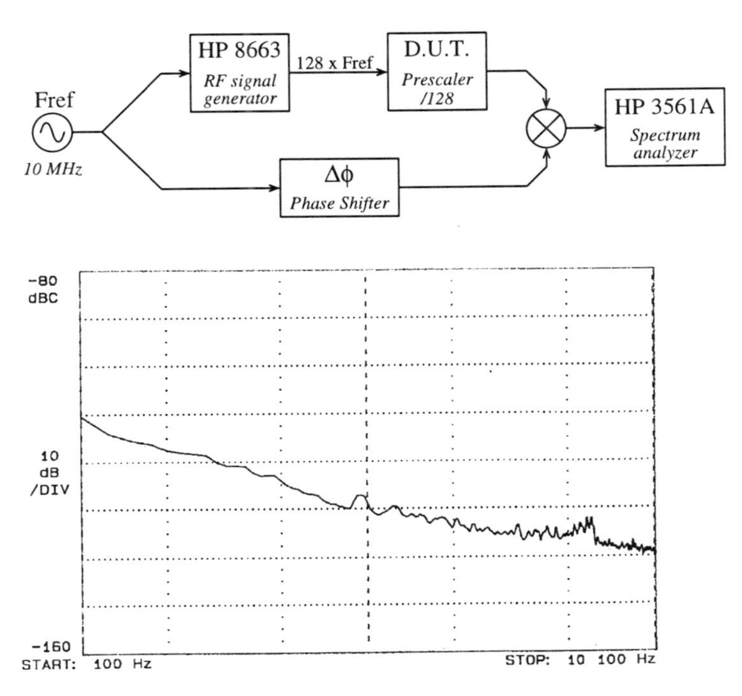

Figure 6.15. Prescaler phase noise measurement: (a) system configuration; (b) result

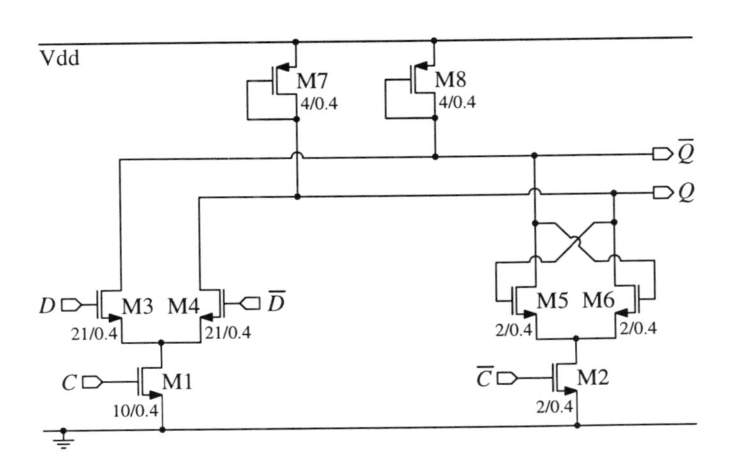

Figure 6.16. Full-speed divide-by-2 flipflop in $0.4\text{-}\mu m$ technology

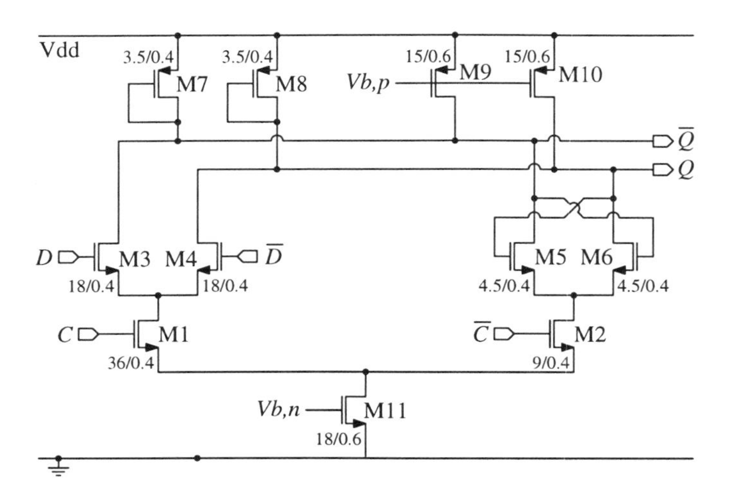

Figure 6.17. Half-speed divide-by-2 flipflop in 0.4-μm technology

Another optimization has been made, i.e. the two parts of the master/slave flipflop have been sized asymmetrically. As can be noted from the general M/S configuration in figure 6.3(a), if one uses only the in-phase outputs (I) to drive the next stage, the slave section is less loaded than the master section. We can therefore size the master section accordingly larger, in order to enable it to drive its load. The transistor sizes given in figure 6.16 are those for the slave, in the master the width of every transistor is a factor 1.67 larger.

For the half-speed divide-by-2, the same modification is possible in this deep sub-micron technology. The load transistors $M7$ and $M8$ are not folded to ground, but are implemented as PMOS diodes to the positive power supply. The current sources $M9$ and $M10$ are not omitted, they are used to make a proper division between the load impedance and the total DC current of the flipflop. The transistor schematic with the sizes indicated is shown in figure 6.17.

The switchable amplifier in the first stage of the phase-selection block can also be simplified by removing the current mirrors. An output stage is added in the form of a small inverter that has its output coupled to its input. That way the overall amplification is of form g_{m1}/g_{m2}, which is set to a value of approximately 7. This extra stage of course reduces the amplification, but it has been added because of two advantages. First, it sets the DC output voltage to the correct value, i.e. the threshold voltage of the next inverter. The transistor sizes have been chosen in order to match that condition. Secondly, because the amplifier output no longer clips to the power supply due to the transients at the switching times, the output settles to its DC value much faster. This

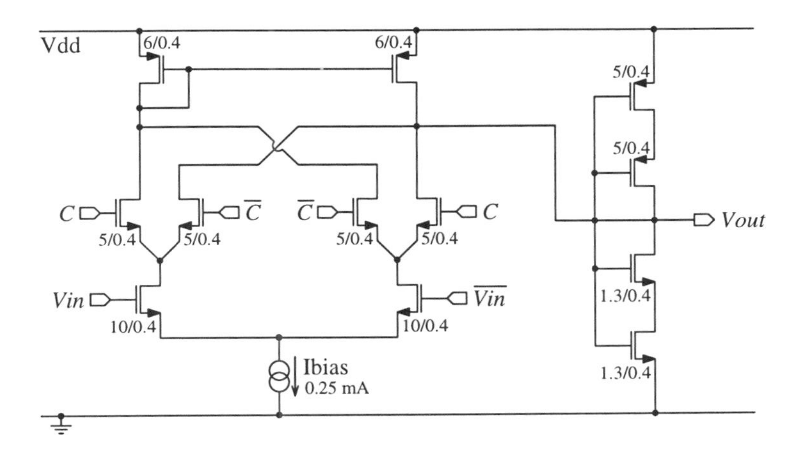

Figure 6.18. Switchable amplifier for phase-selection

allows a much shorter interval in between the switching times, and hence enables to insert more 90° delays per output period. This is a key element in realizing this multi-modulus prescaler. The circuit schematic with transistor sizes is shown in figure 6.18. With respect to the previous design, an inverter is added in between this amplifier and the following NAND gates to further amplify the signals $F.I$ and $F.Q$ before the final selection is made.

The other blocks of the prescaler are similar to the previous design, except of course for the modulus control block. This block has to deliver a number of pulses on its output $Next$, given by the three bits of the division modulus control word $Mode$. The circuit diagram used is shown in figure 6.19(a). It is independent on the relative timing of the signals, which means it will never create spikes at the output if the delay of some gates differs significantly. The timing of the signals is shown in figure 6.19(b). A high level at the $Mode1..4$ control bits enables the corresponding NAND gate to put out a number of pulses equal to 1, 2 or 4 respectively. These pulses are placed such that they do not interfere with each other. The width of the pulses on the signals $Nx1..4$ is determined by the width of the divide-by-8 signal. All pulses are combined in a 3-input NAND gate and an inverter. If the corresponding control bit is set to zero, the $Nx1..4$ signals remain high and no pulse is transferred to the output.

6.4.2 Measurements

This circuit was implemented in a 0.4-μm standard CMOS technology, together with the two VCOs described in section 5.4. Figure 6.20 shows a microphotograph of the 1.8-GHz and the eight-modulus prescaler. The total die area is 1650 × 725 μm^2,

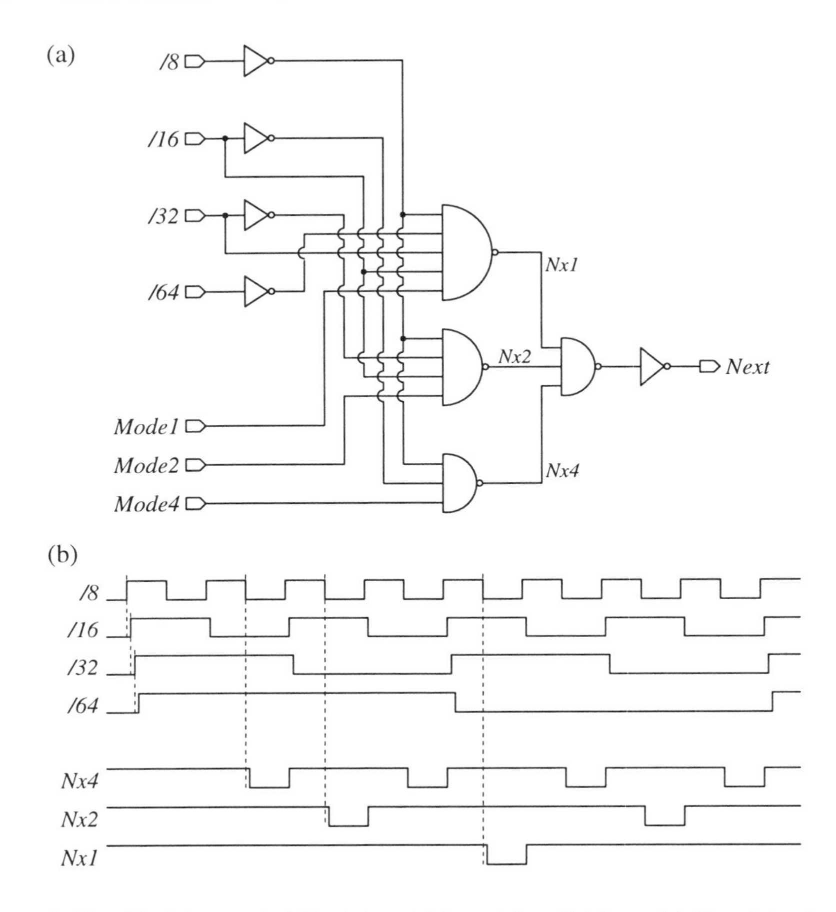

Figure 6.19. Modulus control block for eight-modulus division : (a) Circuit implementation; (b) Waveforms

exactly half of which is occupied by the VCO on the left side of the chip. This VCO is of course very similar to the 0.9-GHz oscillator shown in figure 5.17, except for the coil and some minor changes in transistor and junction capacitor sizes. The prescaler occupies the right hand side of the chip.

The power consumption of the complete circuit is between 5 and 6.5 mA, depending on the division factor. The first divide-by-2 stage takes 1 mA, the second one 0.7 mA. The phase select block and the divide-by-16 consume 1.7 mA and 1.6 mA, respectively. Depending on the division ratio chosen, the power consumption of the modulus control and the phase control blocks varies between 0 and 1.5 mA.

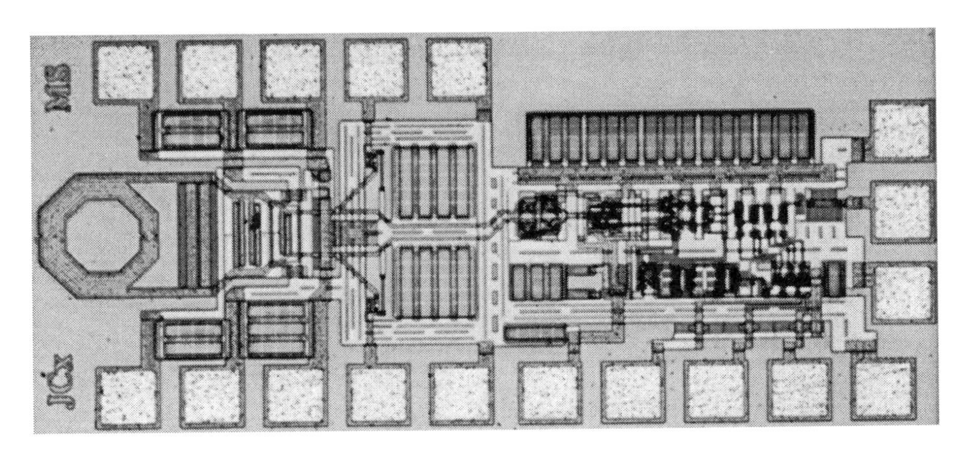

Figure 6.20. Microphotograph of the 0.4-μm CMOS MMP with 1.8-GHz VCO

Figure 6.21 shows a time-domain measurement result of the prescaler operating at high speed. The (highly distorted) output of the VCO at 1.99 GHz is shown, together with the divided output signals for each division modulus. As in figure 6.13, the oscilloscope is triggered by the falling edge of the output signal at time 0 $nsec$. For the divide-by-64 mode the next falling edge arrives after 64 periods of $1/1.99$ GHz, or somewhat later than 32 $nsec$. The output signals for the other moduli are always delayed by an extra period of the VCO signal for each increment in the division number, which can clearly be seen in the figure. A similar chip was also processed, but this time the 900-MHz VCO was used. Also for this input frequency range the prescaler was fully functional, thus indicating that the spike problem of figure 6.11 was properly dealt with.

6.5 CONCLUSIONS

After the improvements made towards the full integration of voltage-controlled oscillators with low-noise specifications in standard CMOS technologies, this chapter has dealt with the other high-frequency building block of a PLL frequency synthesizer, i.e. the frequency divider. Because of the speed limitation of fully programmable counters, most RF designs rely on the use of a dual-modulus prescaler in front of the actual divider, in order to limit its input frequency to a suitable value. But even these DMPs suffer from a speed decrease with respect to a fixed division modulus prescaler. This is because the extra logic added to the flipflops for the dual-modulus operation inherently adds delay to the critical path, and therefore reduces the maximum speed. And the flipflop in a simple divide-by-2 circuit can be optimized for this function, which implies that in a fixed-modulus prescaler faster flipflops can be used. These two ef-

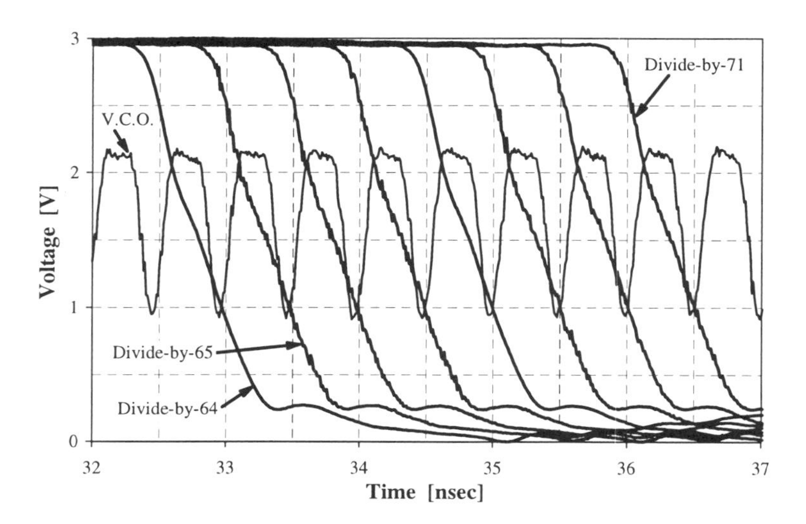

Figure 6.21. Input and output waveforms for all eight division factors at 1.99 GHz

fects add up, and according to [Razav CICC97] the speed ratio between a fixed and a dual-modulus prescaler can be as high as 2 !

We have presented a new architecture for dual-modulus prescalers, the *phase-switching* prescaler, that allows to use a standard divide-by-2 toggle-flipflop in the first stage. Since the speed limit of a prescaler is determined by its first stage, this new prescaler architecture thus allows to achieve the same speeds as an asynchronous fixed-modulus prescaler, but with a dual-modulus functionality. The key element in the phase-switching prescaler is the second divide-by-2 flipflop, which must be a master/slave implementation. Such a M/S flipflop has four outputs which are delayed by 90° with respect to each other. The dual-modulus operating principle is based on periodically switching from one of these signals to the one that is 90° delayed, thus inserting a certain extra delay time to the output. The magnitude of this delay is equal to 90° of a signal at $1/4^{th}$ of the input frequency, or a full period of the input signal. Adding such a delay once per period of the output signal results in an increase in the division modulus by one. Variation on this topology can include a division by $N+1/2$, or a multi-modulus prescaler that divides by all integer number between N and $N+s$, with s an integer larger than or equal to one.

Two designs have been implemented to demonstrate the capabilities of the phase-switching prescaler topology for implementation in a high-frequency PLL LO synthesizer design. The first one is a standalone design in 0.7-μm CMOS. This dual-modulus prescaler achieves a maximum speed of 1.75 GHz with a 3-V power supply. The second one is an eight-modulus prescaler, capable of dividing by all integer number from 64 to 71, and is implemented in standard 0.4-μm CMOS. This design was integrated together with the 1.8-GHz VCO described in chapter 5. It also enables to cover the full frequency band of the DCS-1800 system starting from a 26.6-MHz reference.

Both designs use a CMOS version of an ECL-alike master/slave toggle flipflop in their first stage. The bias current source normally present can be omitted because of the large-swing input signals, which gives another speed enhancement. The circuits are sized to cope with variations in processing parameters, and are indeed the key to the high operation speeds achieved.

Based on the CMOS implementations of the two high-speed building blocks we dispose of now, a completely integrated RF PLL LO synthesizer can be designed with no external components at all. This will be elaborated in the next chapter.

7 A FULLY INTEGRATED CMOS PHASE-LOCKED LOOP FREQUENCY SYNTHESIZER

7.1 INTRODUCTION

Since we now have a high-quality CMOS integrated VCO and a high-speed eight-modulus prescaler, it is possible to realize the ultimate goal of this research, i.e. a complete PLL LO synthesizer for a mobile communication system, integrated in a standard CMOS process without any external components, trimming, or extra processing steps. The DCS-1800 system [ETSI 94] has been taken as a design example.

The design of the VCO and the prescaler has been done with this final goal in mind. They are therefore suited for the most simple PLL block diagram possible. The prescaler division factors (i.e. $64 - 71$) have been chosen in order to use this prescaler as the only block in the PLL frequency divider. Starting from a 26.6-MHz reference, the possible output frequencies range from 1702.4 to 1888.6 MHz, which neatly covers the required frequency band from 1.71 to 1.88 GHz. Of course addition of the fractional-N division technique is necessary to achieve a 200-kHz frequency resolution.

So also no changes have been made to the 0.4-μm CMOS VCO with the hollow spiral inductor discussed in section 5.4, which simply has a tuning range as large as possible to cover the full frequency band of the DCS-1800 system, without the need

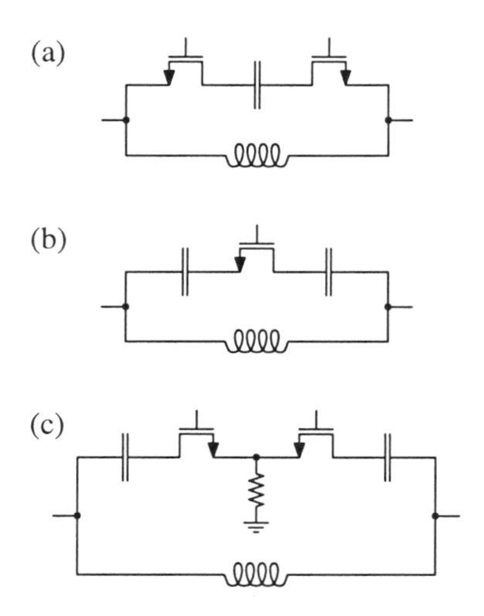

Figure 7.1. Discrete switching of capacitors in a capacitor bank used for coarse tuning

for switching in or out discrete capacitors. This coarse extra tuning that can be added to the normal fine tuning of a VCO increases the frequency range, but the digital logic system that controls the switching must be carefully integrated in the design of the full loop dynamics in order to guarantee correct locking and stability over the complete range. Furthermore, it is not straightforward to add these switched capacitors to the oscillator's LC-tank without deteriorating the overall quality factor. The switches must be implemented as MOS transistors, which always have some series resistance in their on-state, as well as parasitic junction capacitance to the substrate from their source and drain regions. So the circuit design as well as the layout will have an important influence on the resulting quality factor. To get an idea about the order of magnitude of the size of the transistor, we can calculate the required resistance for a 100-fF capacitor. If we want to make a quality factor of 10 at 1.8 GHz, the allowed series resistance is 88 Ω. A transistor operating in its linear region has a conductance equal to $KP \cdot W/L \cdot (V_{GS} - V_T)$, where we can substitute e.g. $KP = 150\ \mu A/V^2$ for the NMOS transistor in the 0.4-μm CMOS process and $V_{GS} - V_T = 2\ V$ in the most optimum case. The resulting W/L equals 40, a value which is not extremely high but nevertheless certainly not negligible. Compared to the W/L values of the amplifying transistors given in section 5.4.2, the switching transistors certainly add a large amount of parasitic capacitance.

Figure 7.1 shows some implementation examples of switched tuning capacitors. The first possibility uses one floating capacitor and two transistor switches. This is not a good example for three reasons. First, two switches means of course twice the resistance. The DC bias point of the circuit sets the drain and source voltage higher than the ground level, e.g. 1 V in the circuits of figures 5.9 and 5.16. The gate-source voltage becomes much less than 3 V and reasonable estimate of $V_{GS} - V_T$ is only 1 V. Finally, as the voltage across the LC-tanks swings significantly, the biasing point of the transistors also varies. One of the transistors will have a lower resistance, but the other one's drain and source voltage will rise very high, resulting in a very low value for $V_{GS} - V_T$ and a high resistance. Because of the increase in the losses at high voltage swings, this effect will probably limit the oscillation amplitude. The source-bulk junction capacitance adds to the amplifier's parasitic capacitance. For a width of 10 μm ($W/L = 40$), this parasitic is approximately 15 fF and of course reduces the tuning range of the VCO's remaining junction capacitor.

The circuit in figure 7.1(b) has only one transistor and two capacitors. This circuit is much better, as the internal point between the two capacitors is a common mode node. The parasitic bottom plate capacitance of the capacitors can be placed on this side, and the biasing of the transistor remains constant. To correctly bias the transistor, a high-impedance resistor to ground can be added at one side. This sets the drain and source voltage to ground, and hence gives the maximum possible value of $V_{GS} - V_T$. However, every capacitor in the switching bank must have its own resistor, and to keep the circuit perfectly symmetrical a resistor should be added at each side of the transistor. This is solved in figure 7.1(c), which uses two transistors at the common mode node. All middle points of all capacitors in the bank can be connected to the same point. Theoretically, the biasing resistor is even not necessary as there is no current flow to the ground if the circuit is perfectly symmetrical. As for the parasitic capacitances, they are not present when the transistor switch is on because they do not have a voltage swing across them. Instead, in the off-state the remaining capacitance is the series circuit of the actual capacitor and the junction capacitance. As this last one is the smallest, it will determine the remaining value. A final problem is the parasitic resistance of the parasitic capacitance, which is not know exactly and very dependent on the substrate doping.

This explains why we have tried to use the basic VCO configuration without switched capacitor bank in the final phase-locked loop block diagram, which is shown in figure 7.2. It is based on a typical 3-rd order type-2 charge-pump PLL, but a lot of improvements have been made in order to achieve full integration of all the components on a single ship. The size of the passive components in the loop filters, especially the capacitors, must be limited by proper design in order to limit the die area.

The following sections will now discuss the implementation aspects of the several building blocks.

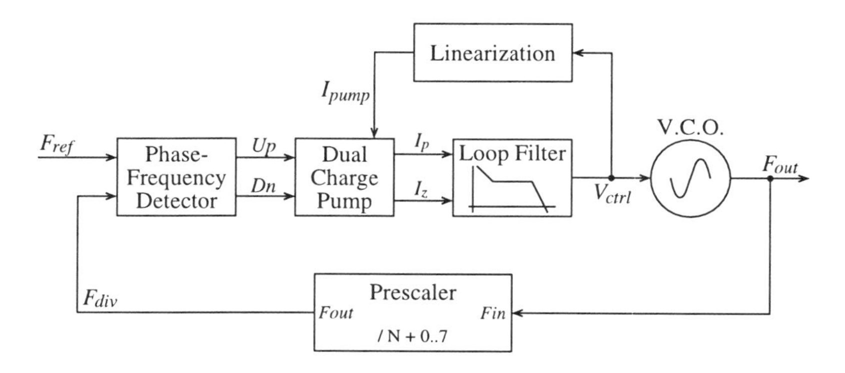

Figure 7.2. Block diagram of the complete $0.4\text{-}\mu m$ PLL frequency synthesizer

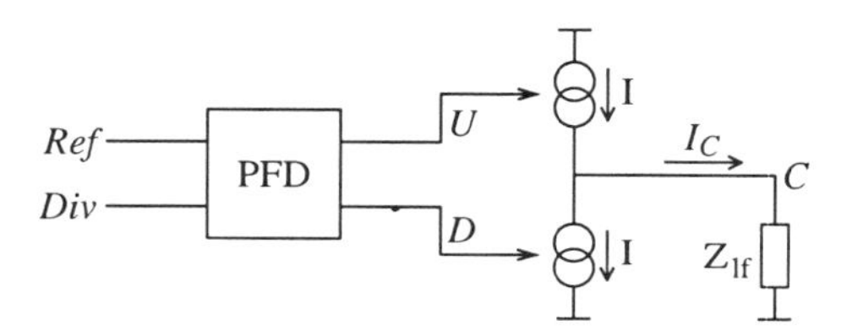

Figure 7.3. Phase-frequency detector with charge pump

7.2 PHASE-FREQUENCY DETECTOR

The phase detector used in this design is a phase-frequency detector. It is used because of its ease in implementation in conjunction with a charge-pump loop filter. A schematic of the combination of the two blocks is shown in figure 7.3 (see also figure 2.10). The PFD schematic used is the crossover-distortion-free circuit already shown in figure 2.11, with the delay block implemented as a string of two inverters. The reference frequency in the PLL block diagram must be 26.6 MHz.

7.3 LOOP FILTER

The loop filter consists of two blocks as shown in figure 7.3, i.e. the charge pump and the filter impedance.

7.3.1 Charge Pump

Basically the charge pump consists of two current sources that are switched on and off at the proper instances in time. A lot of effort has been put into the reduction of

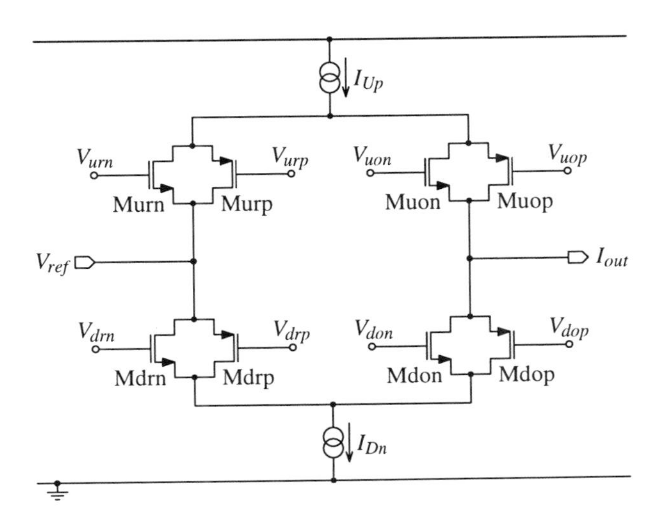

Figure 7.4. Charge pump core circuit with dummy reference branch and two-transistor switches

charge injections at the switching times. Theoretically, a charge-pump PLL does not suffer from reference clock feedthrough once it is locked, because both the *Up* and the *Dn* currents remain off all of the time. But since we have added a delay in the phase-frequency detector to avoid the dead-zone problem, this condition is no longer met. At lock, there will be short pulses on both the *Up* and the *Dn* inputs. Since both currents are of equal amplitude and width, the net effect on the output voltage is zero. But several spikes occur on the output at the times when the currents are switched on and off, which will reflect into the output if no countermeasures are taken. As these spikes occur at the reference frequency, they will cause spurs in the PLL output spectrum at an offset from the carrier equal to this reference frequency.

A transistor schematic of the implemented charge pump core is shown in figure 7.4. Normally, two switches would be enough, one for the up-current I_{Up} and one for the down-current I_{Dn}. In this design, to reduce the reference frequency spurs problem, we have used 8 switches, each with its own control signal.

First, the current sources themselves are not switched on and off, but instead their current is deviated into a reference voltage V_{ref} during the off-state. The switches in the left branch of the circuit are used for this purpose. So either the transistor switches in the left branch are open, and in that case the charge pump output current is zero, or one of the switches in the right branch is open, and in this case the circuit is pumping current up or down. This configuration has the advantage that the response to an *Up*

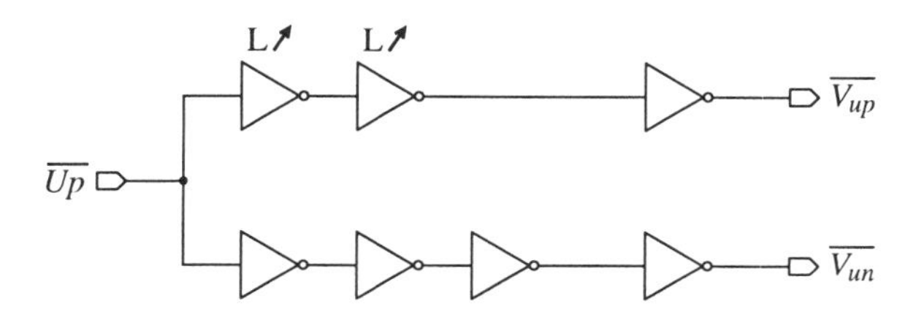

Figure 7.5. Charge pump control signals part 1

or a *Dn* input is immediate, and that we don't suffer from any delays from starting up the current sources. The number of switches required has doubled from 2 to 4.

Secondly, all switches are implemented using both an NMOS and a PMOS switch. This helps in canceling the clock feedthrough of the input pulses to the pump output. Ideally, the control signals of both transistors are of opposite sign and the charge injection cancels.

The naming scheme employed for the eight switches consists of three letters that are chosen as follows. The first letter is a *D* or a *U*, and indicates whether it is a *Down* or an *Up* switch. The second letter indicates the branch the switch is used in, *R* for the dummy path to the reference voltage, and *O* for the signal path to the output node. The last letter is an *N* for the NMOS transistors and a *P* for the PMOS transistors.

There are eight control signals driven these eight transistors. This is again implemented like this in order to reduce the clock feedthrough. We will discuss the signals for the *Up* current as an example. Of course, the signals for the NMOS transistors must be of opposite sign with respect to the signals for the PMOS transistors. Using a simple inverter to derive one signal from the other is not a good solution, since one signal will then always lag the other signal. The charge injection in the switching transistors will also not happen simultaneously and the reference frequency suppression will be insufficient. Both signals must therefore pass through an inverter string, one with an even number of stages and one with an odd number of stages, but both with the same global delay. A circuit diagram is shown in figure 7.5.

The two first inverters in the top path are sized with a larger gate length, such that their delay matches that of the first three inverters in the bottom path, which have minimal sizes. Each path also contains a buffer inverter, which is used to make the rise and fall time of the two signals V_{un} (up, nmos) and V_{up} (up, pmos) equal.

A second circuit tries to adjust the timing of the switches in the left branch in order to overcome some problems seen in the simulations of the charge pump. Theoretically, the left branch must open at the same time when the right branch closes and vice versa. Due to the finite time it requires to actually open and close a transistor switch, in a standard configuration there will always be a short time when both switches are closed, i.e. the opening switch isn't open far enough yet when the closing switch is already completely closed. So the timing of the switches in the reference branch is such that they close after the switch in the output branch opens, and open before the switch in the output branch closes.

The control voltages for this are derived by playing with the threshold voltages of the inverters in a string as shown in figure 7.6. The signal V_{un} is active high, i.e. the switch is closed when the signal is at ground level and is open when the signal is high. The signal V_{uon} will be derived from V_{un}, as well as the signal controlling the up-switch in the reference current branch. Since both signals must be each others complement, the derived reference signal controls the PMOS transistor, which is denoted V_{urp}. So the rising edge of V_{urp} (which closes the switch) must be delayed with respect to the falling edge of V_{uon} (which opens the switch). This is realized by sizing the first inverter in the top path such that it has a threshold voltage that is higher than average while the first inverter in the bottom path must have a threshold voltage that is lower than average. If the input signal goes down, the top path will react faster than the bottom path, as depicted in the second graph of figure 7.6(b). The next inverter in each path is sized just the opposite : a low threshold voltage in the path for V_{uon} and a high threshold voltage in the path for V_{urp}. Every path contains four such inverters, plus a buffering inverter with a normal-threshold inverter to equalize the rise and fall times of the two signals. The resulting signals V_{uon} and V_{urp} are shown in the bottom graph of the figure, and one can easily see that V_{urp} is timed with respect to V_{uon} such that the switch in the reference current branch closes after the switch in the output current branch opens and vice versa for the other direction.

7.3.2 Filter Impedance

Theoretically, one can make a third-order, type two PLL with a passive filter if a charge pump is employed. The pole at zero frequency that is normally formed by an active filter is now replaced by the high-impedance state of the charge pump output. The passive filter requires two capacitors and a resistor as shown in figure 7.7.

As already calculated in section 2.4.2.4, the open loop gain equals

$$\begin{aligned} GH(s) &= \frac{I_{pq} \cdot K_{vco}}{2\pi \cdot N} \times \frac{1 + s \cdot \tau_z}{s^2 \cdot (C_z + C_p) \times [1 + s \cdot \tau_p]} \\ \text{with } \tau_z &= R_z \cdot C_z \\ \tau_p &= R_z \cdot \left(\tfrac{1}{C_z} + \tfrac{1}{C_p}\right)^{-1} \end{aligned} \tag{7.1}$$

The crossover frequency ω_c equals approximately

$$\omega_c \approx \frac{I_{qp} \cdot K_{vco} \cdot R_z}{2\pi \cdot N} \times \frac{C_z}{C_z + C_p} \approx \frac{I_{qp} \cdot K_{vco} \cdot R_z}{2\pi \cdot N} \tag{7.2}$$

We will place the loop gain zero ω_z a factor α below ω_c and the third pole ω_p a factor β above ω_c to guarantee enough phase margin. These factors α and β are

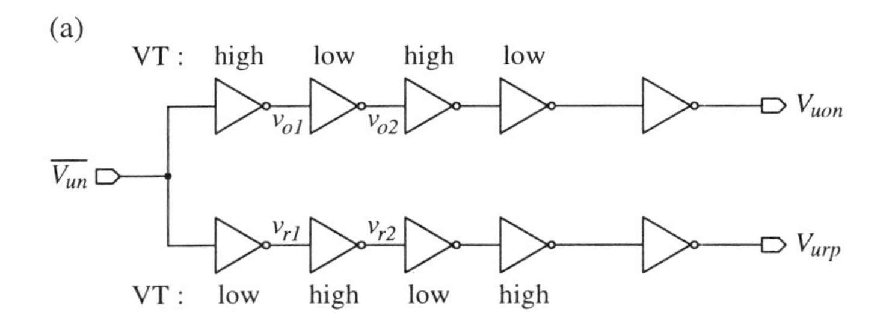

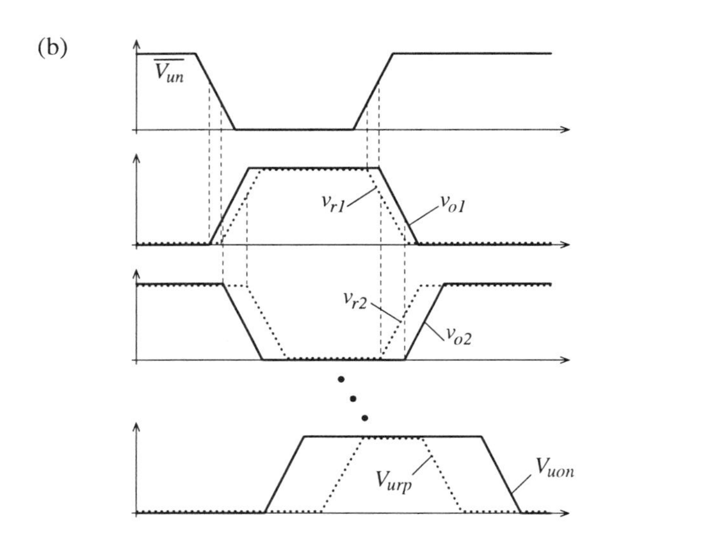

Figure 7.6. Charge pump control signals part 2

typically equal to 4, which gives a phase margin of approximately 60°. The size of the passive elements can easily be calculated from the loop parameters :

$$\begin{aligned} R_z &= \frac{2\pi \cdot N}{I_{qp} \cdot K_{vco}} \cdot \omega_c \\ C_z &= \frac{\alpha}{R_z \cdot \omega_c} = \frac{I_{qp} \cdot K_{vco}}{2\pi \cdot N} \cdot \frac{\alpha}{\omega_c^2} \\ C_p &= \frac{1}{\beta \cdot R_z \cdot \omega_c} = \frac{I_{qp} \cdot K_{vco}}{2\pi \cdot N} \cdot \frac{1}{\beta \cdot \omega_c^2} \end{aligned} \tag{7.3}$$

To make an estimation on the sizes of these passive elements, we can fill in some numbers. Suppose we use a charge pump current of 10 μA, and a crossover frequency

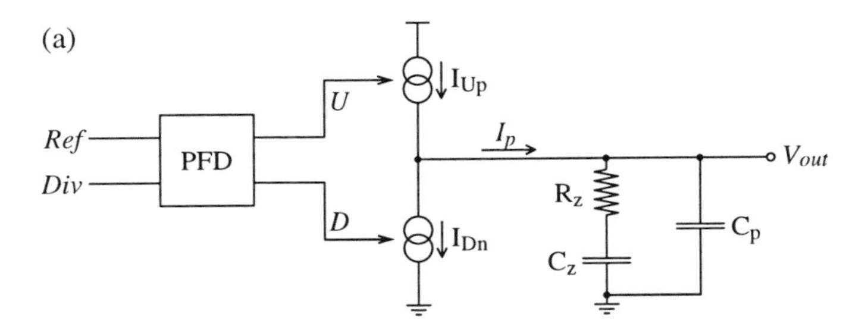

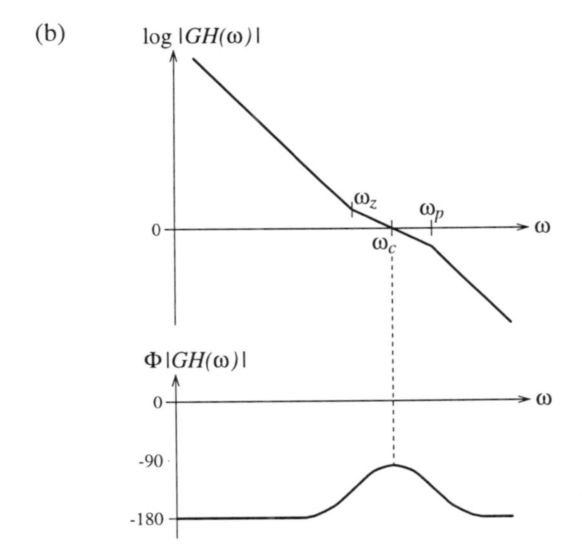

Figure 7.7. (a) Passive loop filter for a 3rd order, type-2 charge pump PLL; (b) Open loop gain Bode plot

of $2\pi \cdot 100\,kHz$. The division modulus N equals 64 and the VCO frequency sensitivity is up to $2\pi \cdot 400\ MHz/V$. This results in $R_z = 10\,k\Omega$, $C_z = 640\ pF$ and $C_p = 40\ pF$.

7.3.3 Active or Passive Filter ?

The combination phase-frequency detector - charge pump - loop filter gives the advantage that a type-2 PLL can be realized without the necessity of an active filter. The use of a passive filter might be easier in the design, but gives some problems in the charge pump itself. In the configuration of figure 7.7, the output voltage of the charge pump varies with the VCO control voltage over a range of approximately 0.7 to 3 V.

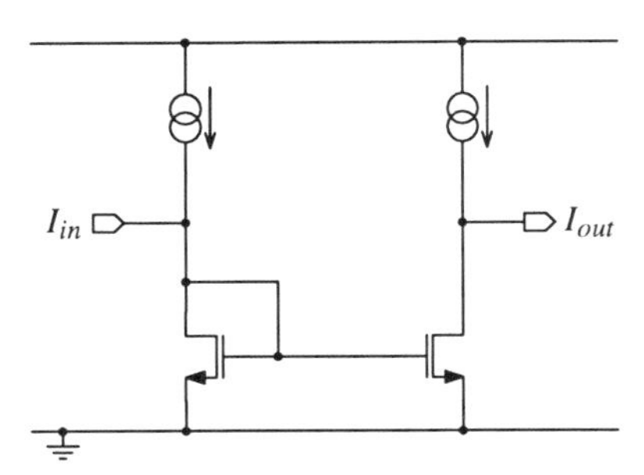

Figure 7.8. Charge pump output current buffer

Under these conditions, the timing of the eight switches can no longer be optimized to reduce the spikes for all output voltages and the reference frequency spurs will not be suppressed. And the finite output impedance of the charge pump gives some other problems.

The magnitude of the *Up* and *Dn* currents of the pump must be matched, otherwise the net charge injected by pulses of equal width into the filter impedance will not be zero. The loop will settle to a small phase offset at the PFD input, resulting in a change in pulse width that compensates the difference in current magnitude. So the net charge injected is zero, but since the current pulses are not equal, reference frequency spikes will again appear at the output.

Another problem that is closely related to this one, is the voltage range that is required at the VCO input, i.e. the voltage must be able to rise as high as possible (preferably up to 3 V) without loosing the saturation condition in the transistors. As will be deducted further on from noise calculations, the $V_{GS} - V_T$ of the charge pump current source transistors should be made as high as possible. These two conditions are not compatible.

A first solution is the insertion of a current buffer at the output of the charge pump, as shown in figure 7.8. If the bias current is much larger than the pump current, the operating point of the NMOS current mirror will not change much, and the output voltage of the pump will remain almost constant. The output current is of the opposite sign of the input current, but this is easily remedied by interchanging the position of the *Up* and the *Dn* inputs.

This current buffer might seem a good idea at first, but some problems remain. The input voltage is dependent on the V_T of the NMOS transistors, which can vary over process conditions. So a V_T-independent biasing must be developed for the current sources. The output impedance of the buffer must also be large, which will certainly require inserting a cascode transistor in the NMOS current mirror and in the PMOS current sources at the top. This again reduces the usable output voltage range. And as will be calculated later, the noise of this buffer will deteriorate the overall noise performance of the PLL.

So despite the charge pump PLL design, we will have to use an active filter.

7.4 NOISE ASPECTS

7.4.1 Charge Pump Noise

The charge pump output current also contains a noise contribution. The current sources of magnitude I_{qp} have a noise current di_n^2. In the locked condition, both current sources are active for a fraction α_{qp} and the current noise adds up quadratically to a value of

$$di_{n,qp}^2 = 2\alpha_{qp} \cdot di_n^2 \tag{7.4}$$

The block diagram that is used to calculate the transfer function from this noise source to the PLL output is shown in figure 7.9. The calculations for this feedback system result in

$$\frac{\theta_{out}}{i_{n,qp}}(s) = \frac{Z_{lf}(s) \cdot K_{vco}}{s + K_{pd} \cdot Z_{lf}(s) \cdot K_{vco} \cdot 1/N} \tag{7.5}$$

For offset frequencies far outside the loop bandwidth, i.e. larger than ω_p, the loop filter impedance can be approximated by $1/sC_p$, and this reduces to

$$\frac{\theta_{out}}{i_{n,qp}}(s) = \frac{2\pi \cdot N}{I_{qp}} \cdot \frac{\beta\omega_c^2}{s^2} \tag{7.6}$$

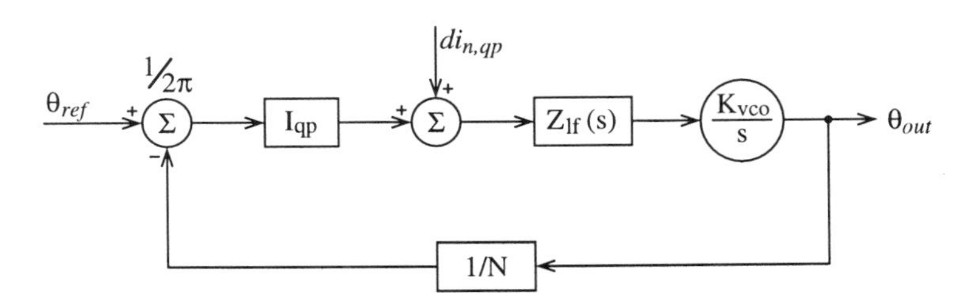

Figure 7.9. Block diagram for the transfer of charge pump noise to output phase noise

If we put $di^2_{n,qp} = 4kT \cdot g_{m,qp} \cdot df$, the single sided spectral phase noise density becomes

$$\begin{aligned}\mathcal{L}_{QP}\{\Delta\omega\} &= \frac{\theta^2_{out}(\Delta\omega)}{2} \\ &= \frac{1}{2} \cdot \left(\frac{2\pi N \cdot \beta\omega_c^2}{I_{qp} \cdot \Delta\omega^2}\right)^2 \times 2\alpha_{qp} \cdot 4kT \cdot g_{m,qp} \\ &= \left(\frac{2\pi N \cdot \beta\omega_c^2}{I_{qp} \cdot \Delta\omega^2}\right)^2 \times \alpha_{qp} \cdot 4kT \cdot \frac{2 \cdot I_{qp}}{(V_{GS} - V_T)_{qp}} \\ &= kT \cdot (4\pi \cdot N \cdot \beta)^2 \cdot \frac{2\alpha_{qp}}{I_{qp} \cdot (V_{GS} - V_T)_{qp}} \cdot \left(\frac{\omega_c}{\Delta\omega}\right)^4\end{aligned} \tag{7.7}$$

Suppose again we use a charge pump current $I_{qp} = 10\ \mu A$. The on-time fraction α_{qp} is typically equal to 0.1, and we can e.g. use a $V_{GS} - V_T$ of 0.5 V. With $\beta = 4$, $N = 64$ and $\omega_c = 2\pi \cdot 100\ kHz$, the single-sided phase noise spectral density at a 600-kHz offset becomes $\mathcal{L}_{QP}\{600\ kHz\} = 1.3\ 10^{-12} =$ -119 dBc/Hz.

This value is 3 dB higher than the phase noise contribution of the VCO. We definitely have to lower this number, in order to achieve the required phase noise spec for the total system. It is our goal to make the sum of the phase noise of all the other components of the PLL smaller than or at least equal to the VCO phase noise, i.e. -122 dBc/Hz.

The charge pump output phase noise $\mathcal{L}_{qp}$ can be lowered a bit by e.g. increasing the $V_{GS} - V_T$ value of the current sources, or decreasing the on-time fraction α_{qp}. This last action will probably increase the spurs in the output spectrum because the timing of the switches will have to be changed. Also the choice of I_{qp} and the loop bandwidth ω_c are very important, not only for the transient characteristics of the PLL, but as demonstrated by the above calculations certainly also for the noise characteristics.

The phase noise caused by a charge pump output current buffer as suggested in figure 7.8 has a similar equation, except that these transistors are always on, so they cannot take advantage of an α_{qp}-factor equal to 0.1. Furthermore, they need a low $V_{GS} - V_T$ value to cover an output voltage range as high as possible. These two aspects will yield very high phase noise numbers, so we can indeed conclude that a current buffer is not feasible due to noise reasons.

7.4.2 Filter Impedance Noise

The resistor R_z in the loop filter impedance also generates white noise $dv^2_{Rz} = 4kt \cdot R_z \cdot df$ that will have a contribution to the output phase noise. A feedback model similar to figure 7.9 can be used for the calculations. For frequencies larger than ω_p, and with R_z expressed as a function of the loop parameters, the phase noise can be approximated by

$$\mathcal{L}_{Rz}\{\Delta\omega\} = kT \cdot (4\pi \cdot N \cdot \beta^2) \cdot \frac{K_{vco}}{I_{qp} \cdot \omega_c} \cdot \left(\frac{\omega_c}{\Delta\omega}\right)^4 \tag{7.8}$$

Using the same numerical values as before, we can evaluate this expression to $\mathcal{L}_{Rz}\{600\ kHz\} = 1.6\ 10^{-11} =$ -108 dBc/Hz. So this value certainly must be reduced in order to achieve the required specs. The most obvious strategy is to lower the loop bandwidth, as there is a dependency on ω_c^3 in the phase noise expression. Obviously, there is a cost involved as the size of the capacitors C_z and C_p are both inversely proportional to the square of ω_c (see equation (7.3)). For example, if we want to decrease the phase noise coming from R_z by 12 dB to -120 dBc/Hz, we must lower the loop bandwidth by a factor $\sqrt[3]{10^{-1.2}} = 2.5$. This increases the value of C_Z by a factor $2.5^2 = 6.3$, or from 640 pF to 4 nF. Also increasing I_{qp} helps according to (7.8), which was also beneficial for the charge pump noise, but again the capacitor values become larger.

These calculations were performed for a passive filter, but the result for an active filter is the same. Of course, the noise contribution from the active elements must also be included. If the filter opamp has an equivalent input voltage noise density of $4kt/G_{mA} \cdot df$, the output phase noise for offset frequencies outside the loop bandwidth equals

$$\mathcal{L}_A\{\Delta\omega\} = 2kT \cdot \frac{K_{vco}^2}{G_{mA} \cdot \Delta\omega^2} \tag{7.9}$$

With a VCO gain constant K_{vco} equal to $2\pi \cdot 400\ MHz/V$, the opamp transconductance needs to be lower than 3.65 mS to achieve a phase noise lower than -120 dBc/Hz at 600 kHz offset.

7.4.3 4-th Order PLL

A solution that reduces the output phase noise of the charge pump and the filter impedance at large offsets is to add another pole in the filter transfer function at the same frequency or a little further than ω_p. This will suppress even more the noise at higher offset frequencies and might allow to relax the sizing of the loop parameters. A straightforward implementation with an active filter is shown in figure 7.10. Of course, an extra noisy element R_4 is introduced, whose influence on the total output phase noise must be considered carefully.

We have placed the extra pole on top of ω_p by making $R_4 \cdot C_4 = \tau_p$. We keep one degree of freedom in doing this, i.e. if we make R_4 a factor γ smaller than R_z, we must increase C_4 by the same amount. To keep the phase margin high enough, β must be increased from 4 to 6. Since the loop transfer function now falls of with 60 dB/dec for frequencies beyond ω_p, the noise of the charge pump and the loop filter resistor

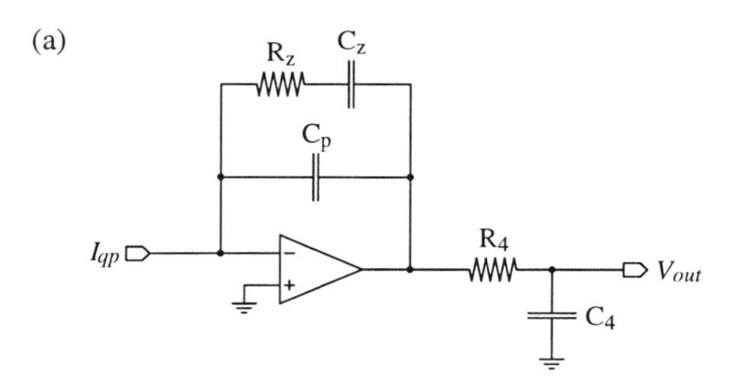

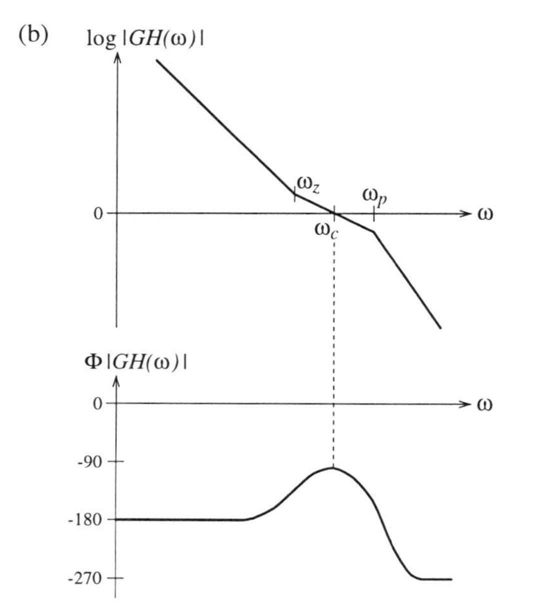

Figure 7.10. (a) Active loop filter for a 4-th order type-2 charge pump PLL; (b) Open loop gain Bode plot

R_z will have a dependency on the offset frequency of $\Delta\omega^{-6}$. The following equations describe this behavior :

$$\mathcal{L}_{QP}\{\Delta\omega\} = kT \cdot (4\pi \cdot N \cdot \beta^2)^2 \cdot \frac{2\alpha_{qp}}{I_{qp} \cdot (V_{GS} - V_T)_{qp}} \cdot \left(\frac{\omega_c}{\Delta\omega}\right)^6 \tag{7.10}$$

$$\mathcal{L}_{Rz}\{\Delta\omega\} = kT \cdot (4\pi \cdot N \cdot \beta^4) \cdot \frac{K_{vco}}{I_{qp} \cdot \omega_c} \cdot \left(\frac{\omega_c}{\Delta\omega}\right)^6 \tag{7.11}$$

$$\mathcal{L}_{A}\{\Delta\omega\} = 2kT \cdot \frac{(\beta \cdot \omega_c \cdot K_{vco})^2}{G_{mA} \cdot \Delta\omega^4} \tag{7.12}$$

$$\mathcal{L}_{R4}\{\Delta\omega\} = kT \cdot (4\pi \cdot N \cdot \beta^2) \cdot \frac{K_{vco}}{\gamma \cdot I_{qp} \cdot \omega_c} \cdot \left(\frac{\omega_c}{\Delta\omega}\right)^4 \tag{7.13}$$

The first three equations are similar to those for a third-order loop. There is an improvement in phase noise with a factor $(\beta\omega_c/\Delta\omega)^2$. This result could be expected, as this is exactly the extra suppression we get from the fourth pole. The phase noise contribution of R_4 is normally larger than that from R_z, but is can easily be made small be choosing a large value for γ. This however also leads to a large value for C_4 and hence once again a large chip area.

7.5 IMPROVED LOOP FILTER

As can be deducted from the numerical examples in the previous paragraphs, it will be difficult to realize a phase noise low enough without the need for several nF of filter capacitance. The insertion of a fourth pole helps to reduce the phase noise at high offsets, but the only thing that really helps is to reduce the loop bandwidth. A lower loop bandwidth will suppress all the noise contributions from the components in the loop, i.e. the PFD, the charge pump, the loop filter, and also those from the reference signal and from the frequency divider. We do not want to decrease the loop bandwidth to very small values, as in that case the transient response of the loop when changing from one channel to another will deteriorate.

The DCS-1800 system we use as a target in our design, is a time-multiplexed system with 8 time slots. Each time slot is 577 μsec wide and is used as shown in

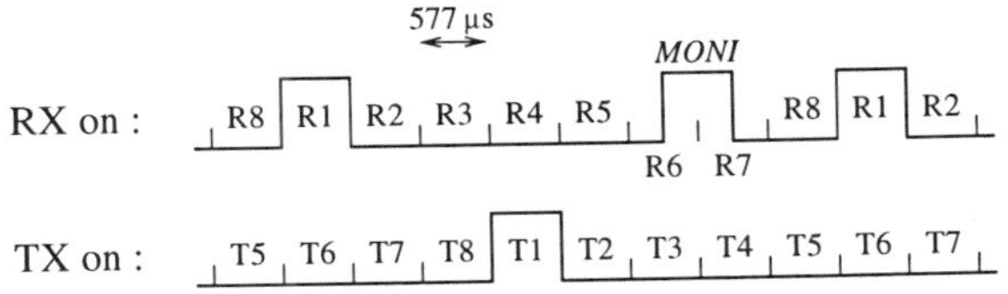

Figure 7.11. DCS-1800 receive (RX) and transmit (TX) time slots

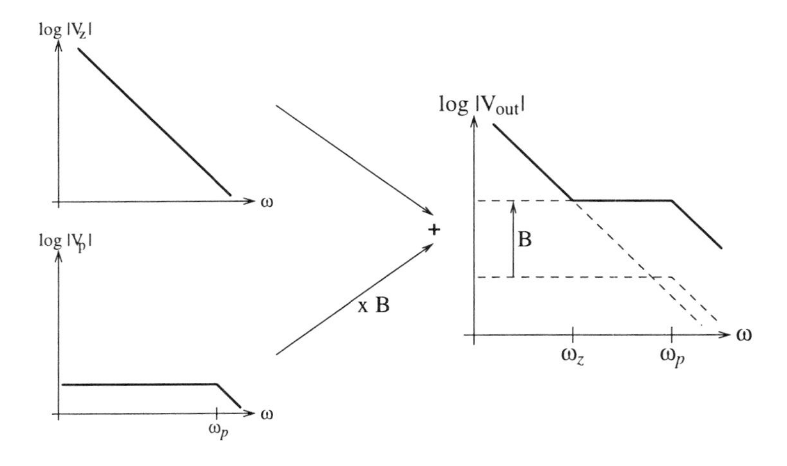

Figure 7.12. Dual-path loop filter principle

figure 7.11. The mobile terminal receives data in slot $R1$ and transmits during $T1$ two time slots later. Before the following receive slot a monitor slot is inserted for system operation. This slot is positioned in between $R6$ and $R7$. So the worst case situation for settling is when the synthesizer has to switch from the transmit band to the receive band for monitoring in one and a half time slot, or 865 μsec. To include some safety margin, we could specify a minimum required PLL settling time of 700 μsec. From equation (2.27), the time required to settle a first-order loop to a new output frequency within an accuracy ε equals -ln ε/ω_c. For a frequency step in the order of 100 MHz, and a final accuracy of 100 Hz ($\varepsilon = 10^{-6}$), the minimum loop bandwidth required in this case is 3.1 kHz. Of course, the settling time for a third or fourth order loop cannot be described by a simple formula like this, so the actual settling time must be obtained be simulation.

7.5.1 Filter topology

When the loop bandwidth is decreased, countermeasures must be taken against the enormous increase in capacitance value. The solution to this is found in splitting up the loop filter, as was already done in [Mijus JSSC94]. The principle is explained in figure 7.12. The passive loop filter consisting of R_z, C_z and C_p actually forms a pole at zero frequency, a zero at ω_z and another pole at ω_p. It is the capacitor creating the zero that has the largest size and hence poses difficulties in integration. In this new technique the same filter characteristic is achieved by combining two signals, that does not need an actual RC combination at the frequency of the zero.

The goal is to add two signals, lets call them V_z and V_p. The first signal is an integrated version of the input current. So it has a pole at zero frequency :

$$V_z = \frac{1}{sC_z} \cdot I_{in} \tag{7.14}$$

The other signal path has a low-pass transfer function :

$$V_p = \frac{R_p}{1 + sR_pC_p} \cdot I_{in} \tag{7.15}$$

These two signals are now added to form the complete output signal, but the second one is amplified by a factor B :

$$\begin{aligned} V_{out} &= V_z + B \cdot V_p \\ &= \left[\frac{1}{sC_z} + B \cdot \frac{R_p}{1 + sR_pC_p}\right] \cdot I_{in} \\ &= \frac{1}{sC_z} \cdot \frac{1 + sR_pC_p + sBR_pC_z}{1 + sR_pC_p} \cdot I_{in} \\ &= \frac{1}{sC_z} \cdot \frac{1 + s\tau_z}{1 + s\tau_p} \cdot I_{in} \\ &\text{with } \tau_z = R_p \cdot (C_p + BC_z) \approx BR_pC_z \\ &\quad\;\; \tau_p = R_pC_p \end{aligned} \tag{7.16}$$

So a large time constant (or a low frequency) is realized for the filter zero, without the requirement for a large capacitor, as τ_z profits from the multiplication by the factor B. The implementation for this configuration is shown in figure 7.13. The circuit has two input currents, one to be integrated to V_z and one to be low-passed to V_p. The impact on the charge pump design is minimal, as only the charge pump core must be implemented twice. The control signals can be re-used for both parts. And multiplication of the second signal by a factor B is also very easy to do, as this is done in the current domain by using a pump current that is a factor B larger in the charge pump core driving the signal V_p. Both input currents are here represented with a negative sign, which means that the position of the Up and the Dn terminals must be interchanged in the phase-frequency detector. The adder that sums the two signals is actually a subtracter the subtracts -$B \cdot V_p$ from V_z. Finally, a fourth pole is added by the combination of R_4 and C_4.

Moreover, in the configuration of figure 7.13, the DC operating voltage of both current inputs is positioned at V_{ref}. This is obvious in the active integrater, but it is also true for the low-pass path as no current flows through R_p when the loop is in lock. So an active implementation for the R_p-C_p path is not necessary.

The circuit implementation proposed for the filter adder is shown in figure 7.14. The transistors controlled by a biasing voltage $V_{b,n}$ or $V_{b,p}$ act as current sources. The

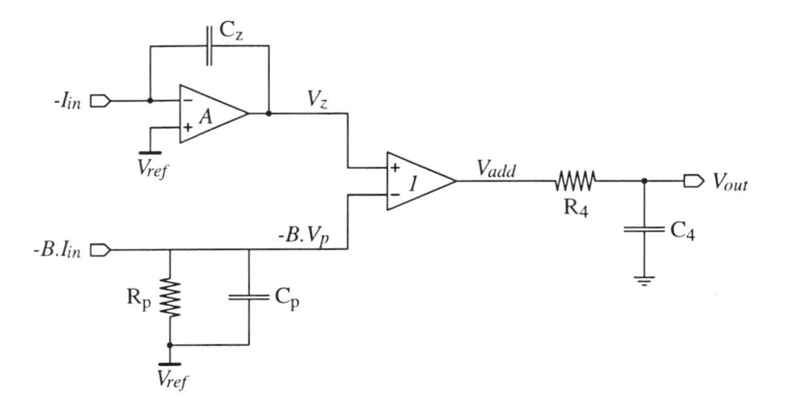

Figure 7.13. Dual-path loop filter implementation

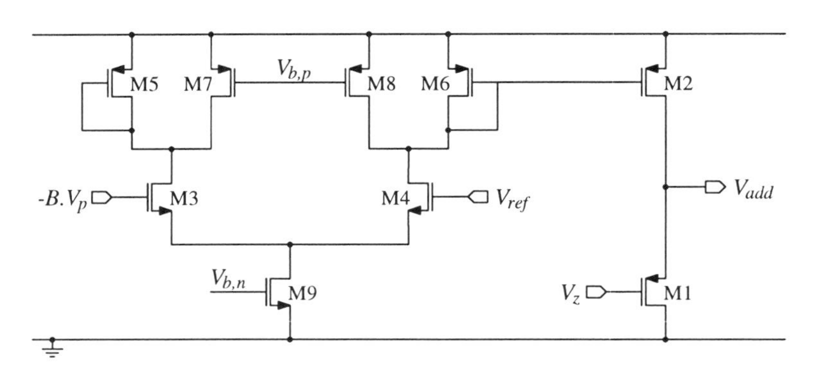

Figure 7.14. Dual-path loop filter adder circuit schematic

signal V_z must be transferred to the output with a positive sign, which is done using a PMOS source follower. The output voltage must be able to fall as low as possible to cover the complete possible tuning range of the VCO. Therefore this output structure is chosen. The bulk of transistor $M1$ is connected to its source (this is possible for a PMOS transistor in an N-well process) to remove the threshold voltage increase due to the bulk effect, and a low $V_{GS} - V_T$ is chosen ($0.125\ V$). With a threshold voltage of $0.65\ V$, the gate-source voltage of $M1$ equals $0.775\ V$.

The transistor $M2$ transfers the signal V_p to the output. Its $V_{GS} - V_T$ is chosen rather low (i.e. $0.25\ V$) for the same reason of the VCO tuning range. A differential pair with the other terminal connected to V_{ref} is the best suited input stage, as the DC operating point of V_p will always be at that voltage. The output transfer function

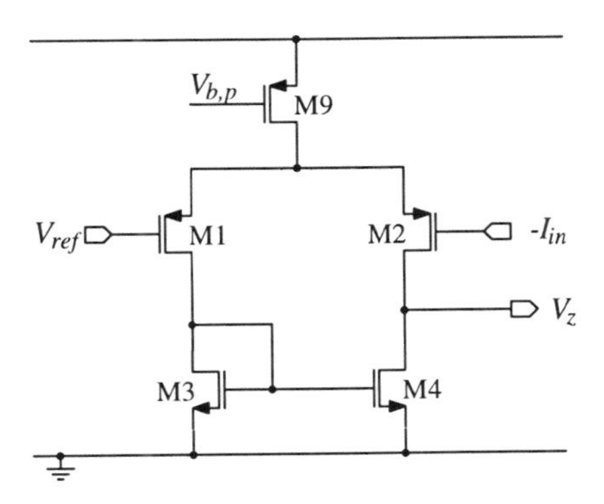

Figure 7.15. Dual-path loop filter opamp circuit schematic

equals

$$V_{add} = \left(-\frac{g_{m2}}{g_{m1}} \cdot \frac{g_{m4}}{2 \cdot g_{m6}} \right) \times (-B \cdot V_p) \tag{7.17}$$

The transconductances are sized such that this transfer function equals one.

The transistor schematic for the integrater opamp is shown in figure 7.15. A PMOS input differential pair is chosen in order to be able to achieve output voltages as low as possible. It is this output voltage that will drive the PMOS source follower of the adder in figure 7.14. Transistors $M3$ and $M4$ have a very low $V_{GS} - V_T$ value of 0.125 V, which makes the lowest possible tuning voltage of the VCO equal to $0.775 + 0.125 = 0.9\ V$.

7.5.2 Transfer Functions

We will calculate the open loop gain and the several noise transfer functions of the PLL with this improved loop filter.

7.5.2.1 Open Loop Gain. The loop filter impedance of figure 7.13 is calculated as follows :

$$\begin{aligned}
V_{out} &= V_{add} \times \frac{R_4}{1 + s \cdot R_4 \cdot C_4} \\
&= [V_z + B \cdot V_p] \times \frac{R_4}{1 + s \cdot R_4 \cdot C_4} \\
&= \left[\left(\frac{-1}{s \cdot C_z} \cdot -I_{in}\right) - \left(B \cdot \frac{R_p}{1 + s \cdot R_p \cdot C_p} \cdot -I_{in}\right)\right] \times \frac{R_4}{1 + s \cdot R_4 \cdot C_4} \\
&= \frac{1}{s \cdot C_z} \cdot \frac{1 + s \cdot \tau_z}{(1 + s \cdot \tau_p)(1 + s \cdot \tau_4)} \cdot I_{in} \\
\text{with } \tau_z &= R_p \cdot (C_p + BC_z) \approx BR_pC_z \\
\tau_p &= R_pC_p \\
\tau_4 &= R_4C_4
\end{aligned} \tag{7.18}$$

This gives an open loop gain equal to :

$$GH(s) = \frac{I_{qp} \cdot K_{vco}}{2\pi \cdot N} \times \frac{1 + s \cdot \tau_z}{s^2 \cdot C_z \times (1 + s \cdot \tau_p) \cdot (1 + s \cdot \tau_4)} \tag{7.19}$$

The crossover frequency is

$$\omega_c \approx \frac{I_{qp} \cdot K_{vco}}{2\pi \cdot N} \times \frac{C_p + B.C_z}{C_z} \approx \frac{I_{qp} \cdot K_{vco} \cdot B.R_p}{2\pi \cdot N} \tag{7.20}$$

provided that C_p can be neglected with respect to $B \cdot C_z$.

We will put the zero $\omega_z = 1/\tau_z$ a factor α below the loop bandwidth. Considering the two high-frequency poles $1/\tau_p$ and $1/\tau_4$, the best results for noise suppression outside the loop bandwidth versus phase margin are obtained if they coincide. We will place them a factor β above ω_c. So if the resistor R_4 is made a factor γ smaller than R_p, C_4 must be larger than C_p by the same amount. This gives the following equations for the passive element values :

$$\begin{aligned}
R_p &= \frac{1}{B} \cdot \frac{2\pi \cdot N}{I_{qp} \cdot K_{vco}} \cdot \omega_c \\
C_z &= \frac{\alpha}{B \cdot R_p \cdot \omega_c} = \frac{I_{qp} \cdot K_{vco}}{2\pi \cdot N} \cdot \frac{\alpha}{\omega_c^2} \\
C_p &= \frac{1}{\beta \cdot R_p \cdot \omega_c} = \frac{I_{qp} \cdot K_{vco}}{2\pi \cdot N} \cdot \frac{B}{\beta \cdot \omega_c^2} \\
R_4 &= \frac{R_p}{\gamma} \\
C_4 &= \gamma \cdot C_p
\end{aligned} \tag{7.21}$$

7.5.2.2 Charge Pump Noise. The noise contribution to the output phase noise from the charge pump current sources can again be calculated. For frequencies above the third and fourth pole, the transfer function from the first noisy charge pump to the output is approximated by :

$$\frac{\theta_{out}}{i_{n,qp_z}}(s) = \frac{2\pi \cdot N \cdot \beta}{I_{qp} \cdot \alpha} \cdot \left(\frac{\omega_c}{s}\right)^3 \tag{7.22}$$

and for the second charge pump the result equals

$$\frac{\theta_{out}}{i_{n,qp_p}}(s) = \frac{2\pi \cdot N \cdot \beta^2}{I_{qp} \cdot B} \cdot \left(\frac{\omega_c}{s}\right)^3 \tag{7.23}$$

The noise magnitudes are given by

$$\begin{aligned} di^2_{n,qp_z} &= 2\alpha_{qp} \cdot 4kT \cdot g_{m,qp_z} \cdot df \\ di^2_{n,qp_p} &= 2\alpha_{qp} \cdot 4kT \cdot g_{m,qp_p} \cdot df = 2\alpha_{qp} \cdot 4kT \cdot B \cdot g_{m,qp_z} \cdot df \end{aligned} \tag{7.24}$$

and both contributions must be added quadratically :

$$\begin{aligned} \theta^2_{out}(s) &= \left[\frac{2\pi \cdot N \cdot \beta}{I_{qp} \cdot \alpha} \cdot \left(\frac{\omega_c}{s}\right)^3\right]^2 \times 2\alpha_{qp} \cdot 4kT \cdot g_{m,qp_z} \\ &\quad + \left[\frac{2\pi \cdot N \cdot \beta^2}{I_{qp} \cdot B} \cdot \left(\frac{\omega_c}{s}\right)^3\right]^2 \times 2\alpha_{qp} \cdot 4kT \cdot B \cdot g_{m,qp_z} \\ &= \left(\frac{2\pi \cdot N \cdot \beta}{I_{qp}}\right)^2 \cdot \left(\frac{1}{\alpha^2} + \frac{\beta^2}{B}\right) \cdot 2\alpha_{qp} \cdot 4kT \cdot g_{m,qp_z} \cdot \left(\frac{\omega_c}{s}\right)^6 \end{aligned} \tag{7.25}$$

Substituting $g_m = 2I/(V_{GS} - V_T)$, the phase noise becomes

$$\begin{aligned} \mathcal{L}_{QP}\{\Delta\omega\} &= \frac{\theta^2_{out}(\Delta\omega)}{2} \\ &= kT \cdot (4\pi \cdot N \cdot \beta)^2 \cdot \frac{2\alpha_{qp}}{I_{qp} \cdot (V_{GS} - V_T)_{qp}} \cdot \left(\frac{1}{\alpha^2} + \frac{\beta^2}{B}\right) \cdot \left(\frac{\omega_c}{\Delta\omega}\right)^6 \end{aligned} \tag{7.26}$$

Comparing this to the result (7.10), we see that for the same pump current I_{qp} the normal 4th-order loop filter implementation has a phase noise contribution that is a factor

$$\frac{\beta^2}{\frac{1}{\alpha^2} + \frac{\beta^2}{B}} = \frac{\alpha^2 \beta^2 B}{\alpha^2 \beta^2 + B} \approx B \tag{7.27}$$

larger than this improved loop filter implementation.

7.5.2.3 Loop Filter Noise. The advantage of this improved loop filter implementation also shows up in the phase noise contribution of the resistor R_p. Again, the phase noise contributions at offset frequencies larger than $\beta\omega_c$ can be approximated with :

$$\mathcal{L}_{Rp}\{\Delta\omega\} = kT \cdot \frac{4\pi \cdot N \cdot \beta^4 \cdot K_{vco}}{B \cdot I_{qp} \cdot \omega_c} \cdot \left(\frac{\omega_c}{\Delta\omega}\right)^6 \tag{7.28}$$

As could be expected since the value of R_p can be made a factor B smaller than R_z in the normal filter implementation, the phase noise contribution of R_p is also a factor B smaller, which is an advantage that cannot be underestimated.

The other elements in the loop filter cause the following amounts of output phase noise :

$$\mathcal{L}_A\{\Delta\omega\} = 2kT \cdot \frac{(\beta \cdot \omega_c \cdot K_{vco})^2}{G_{mA} \cdot \Delta\omega^4} \tag{7.29}$$

$$\mathcal{L}_D\{\Delta\omega\} = 2kT \cdot \frac{(\beta \cdot \omega_c \cdot K_{vco})^2}{G_{mD} \cdot \Delta\omega^4} \tag{7.30}$$

$$\mathcal{L}_{R4}\{\Delta\omega\} = kT \cdot \frac{4\pi \cdot N \cdot \beta^4 \cdot K_{vco}}{B \cdot \gamma \cdot I_{qp} \cdot \omega_c} \cdot \left(\frac{\omega_c}{\Delta\omega}\right)^4 \tag{7.31}$$

Here $\mathcal{L}_A$ represents the noise coming from the amplifier in the integrater with C_z, which is supposed to have an equivalent input voltage noise of $4kT/G_{mA}$. Its magnitude is the same as in the normal filter implementation. An extra active element has been added in this new configuration, which also adds to the output phase noise. With an equivalent input voltage noise of $4kT/G_{mD}$, the phase noise contribution $\mathcal{L}_D$ has the same shape as $\mathcal{L}_A$. Finally, the resistor R_4 has a value that is a factor B smaller than in the normal filter, and hence its phase noise is lowered with respect to (7.13) by the same amount.

7.5.3 Filter Optimization

We can now decrease the loop bandwidth to achieve the phase noise specs, and find the optimum loop parameters. This was done using a behavioral linear model for the complete PLL that was implemented in HSpice [HSpice96] as shown in figure 7.16. Normally, a behavioral PLL model uses the phase of the signal as a state variable, e.g. θ_{ref} for the input signal. In this Spice model these phases are represented by a voltage such as $V_{\theta ref}$.

The phase detector is a voltage-controlled voltage source with a gain E_{pd} equal to $1/2\pi$. The resulting phase error is converted into a current in the two charge pumps using a voltage-controlled current source. As the top charge pump current must be I_{qp}, its gain $G_{qp,z}$ equals I_{qp} $[A/V]$. The bottom charge pump has a gain $G_{qp,p} = B \cdot I_{qp}$ $[A/V]$. The loop filter is implemented in the voltage domain, of course. Either

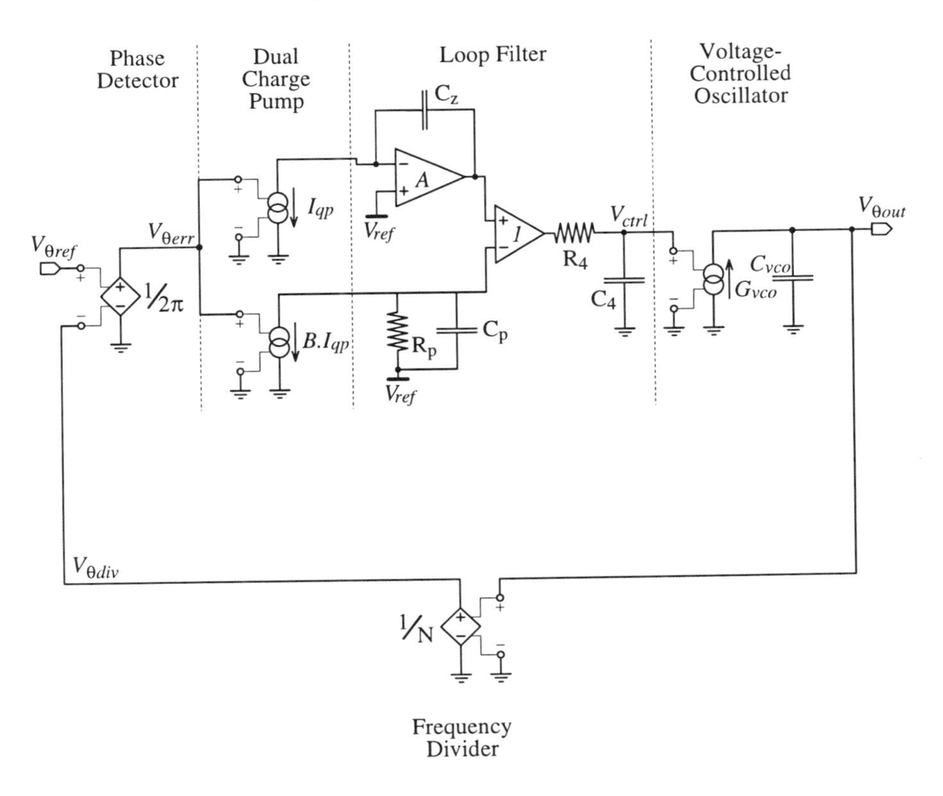

Figure 7.16. Behavioral PLL model implemented in Spice

ideal circuit blocks can be used, or the transistor schematics of figures 7.14 and 7.15 can be implemented in this model. To implement the VCO transfer characteristic $\theta_{out} = K_{vco}/s \cdot V_{ctrl}$ we use a voltage controlled current source G_{vco} with a capacitive load C_{vco}. This gives a transfer function

$$V_{\theta out} = \frac{1}{s \cdot C_{vco}} \cdot G_{vco} \cdot V_{ctrl} \tag{7.32}$$

So the capacitor and the transconductor value must be chosen such that $G_{vco}/C_{vco} = K_{vco}$. The frequency divider is implemented as a voltage-controlled voltage source transferring $V_{\theta out}$ to $V_{\theta div}$ with a gain $E_{div} = 1/N$. Some biasing elements that are required to set all the correct operating points are not indicated in the figure. The appropriate noise sources can be introduced, and the resulting output phase noise is calculated with

$$\mathcal{L}\{\Delta\omega\} = \frac{S_{V_{\theta out}}\{\Delta\omega\}}{2} \tag{7.33}$$

Using this model, the loop parameters can be varied and their effect on the PLL performance can be evaluated rapidly. We cannot change the relative position of the zero $1/\tau_z$ and the two poles at $1/\tau_p$ because the phase margin must be high enough in order to maintain a stable system. So we put $\alpha = 4$ and $\beta = 6$. The VCO gain K_{vco} equals at most $2\pi \cdot 400MHz/V$ and the division modulus N is 64. So the remaining loop parameters that are incorporated into the optimization are the loop bandwidth ω_c, the charge pump current I_{qp}, the filter factor B and the fourth pole parameter γ. Our goal is to achieve a low phase noise within two constraints. The first constraint is a low power consumption, which will be reflected in the restrictions on the values of G_{mA} and G_{mD}. The second constraint is the most important one and limits the occupied chip area. We must limit the sum of the capacitor sizes of C_z, C_p and C_4 to a value realizable in an integrated circuit.

Generally, the basic way of determining all the loop parameters can begin with setting the charge pump current as small as possible, since this will lead to the smallest capacitor values according to 7.21. Unfortunately, this cannot be done down to very low currents as the output phase noise increases. Not only the phase noise outside the loop bandwidth, as has been calculated in the previous sections, is important, but also inside the loop bandwidth the phase noise must remain below the level of -80 dBc/Hz according to the specification for the DCS-1800 system. This cannot be captured in a simple formula, but it is easily seen in the simulations. Also some practical reasons limit the minimum charge pump current, such as the charge injection of the transistor switches in the pump. For the same minimum transistor sizes, the amount of injected charge remains constant and becomes relatively more and more important as the pump current decreases. As a consequence, the spurs in the output spectrum at the reference frequency offset will increase. With this minimum pump current, the loop bandwidth can be set to a value that allows to achieve the out-of-band phase noise specs. Of course, a few iterations of this process might be required in order to achieve the optimal performance.

The results of the final optimized circuit are shown in figures 7.17 and 7.18. The loop parameters are listed in table 7.1. A pump current of 1 μA is used, and the loop bandwidth equals 45 kHz. This should be sufficiently large for the settling time. The total capacitance value is as large as 960 pF. This will be responsible for a large chip area, but it is the only way to achieve the required low phase noise. Even with these values, the spec is only achieved without any margin, i.e. any other phase noise introduced in the circuit will cause the phase noise to be higher than allowed in the DCS-1800 system. If the VCO phase noise at 600 kHz, i.e. -122.5 dBc/Hz is added, the total PLL output phase noise rises to -117 dBc/Hz. However, this situation only appears at the highest possible VCO gain, or at the minimum tuning

Loop bandwidth	ω_c	45 kHz
Charge pump current	I_{qp}	1 μA
Zero frequency	α	4
Pole frequency	β	6
Filter current ratio	B	12
Fourth pole ratio	γ	3
Passive elements :	C_z	320 pF
	R_p	3.6 $k\Omega$
	C_p	160 pF
	R_4	1.2 $k\Omega$
	C_4	480 pF
Phase noise contributions at 600 kHz:		
Charge pump	$\mathcal{L}_{QP}$	-137 dBc/Hz
Resistor R_p	$\mathcal{L}_{Rp}$	-124 dBc/Hz
Integrater opamp	$\mathcal{L}_A$	-129 dBc/Hz
Filter adder	$\mathcal{L}_D$	-129 dBc/Hz
Resistor R_4	$\mathcal{L}_{R4}$	-121 dBc/Hz
Total	$\mathcal{L}\{600\ kHz\}$	-118.6 dBc/Hz

Table 7.1. Final PLL parameters

voltage. In the suggested loop filter integrater and adder implementation, this voltage cannot drop below 0.9 V, where the VCO gain has already fallen to 250 MHz/V. As can be deducted from the equations (7.26) through (7.31), this evolution will have a beneficial effect on the phase noise. This will be discussed further in the section on the linearization of the VCO.

Figure 7.17 shows the simulated open-loop gain, which has indeed a crossover frequency of 45 kHz. The phase margin is 58°, a value which is high enough to avoid excessive ringing and slow settling. In the charge pump, an on-time fraction α_{qp} of 0.1 and a $V_{GS} - V_T$ of 0.3 V has been used. Figure 7.18 shows the several phase noise contributions versus frequency for the designed loop parameters. We can clearly see the dependency on $\Delta\omega^{-4}$ or $\Delta\omega^{-6}$ for high offset frequencies. The values at 600 kHz

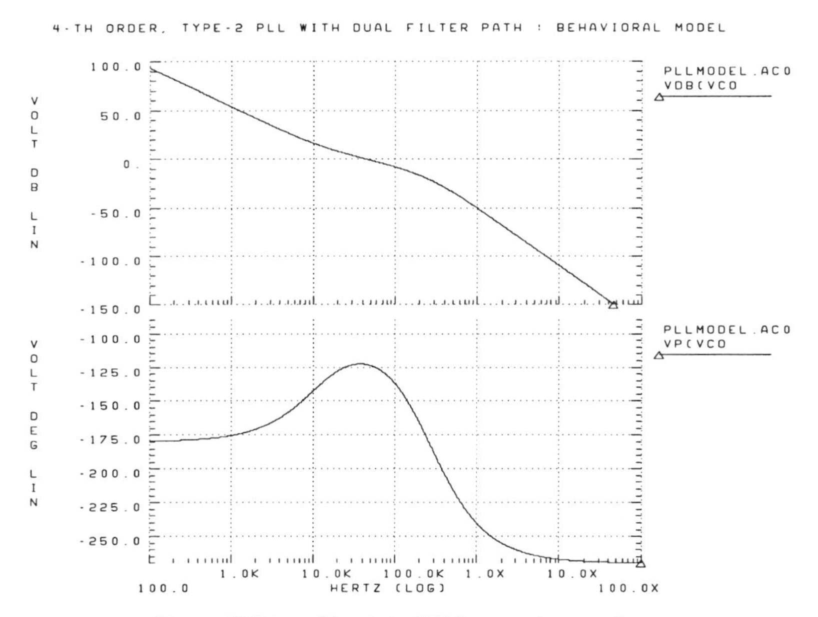

Figure 7.17. Simulated PLL open-loop gain

were already listed in the above table, and they agree very well with the theoretical formulas (7.26) through (7.31).

7.6 LINEARIZATION

As known from the design and measurements of the VCO, the tuning curve of the VCO is highly non-linear. A variation in K_{vco} will change the open-loop gain and hence change the loop bandwidth and the phase margin. A small deviation can be tolerated, but soon the PLL feedback loop will become unstable. It is therefore necessary to linearize the VCO sensitivity.

A first and most obvious solution for this is the insertion of a circuit block in between the loop filter and the VCO that compensates for the non-linear VCO gain. This is shown in figure 7.19(a). The input control voltage V_{ctrl} is not applied directly to the VCO, but is converted to another voltage V_{tune} with a non-linear characteristic. The product of this transfer curve with the VCO tuning sensitivity should be as constant as possible to achieve a linear overall tuning.

A circuit that approximately performs this compensation is shown in figure 7.19(b). It is based on the quadratic MOS I/V characteristic. At low input voltages, the current through and hence also the transconductance of the input transistor $M1$ is low. The output voltage V_{tune} changes slowly with the input voltage V_{ctrl}, which compen-

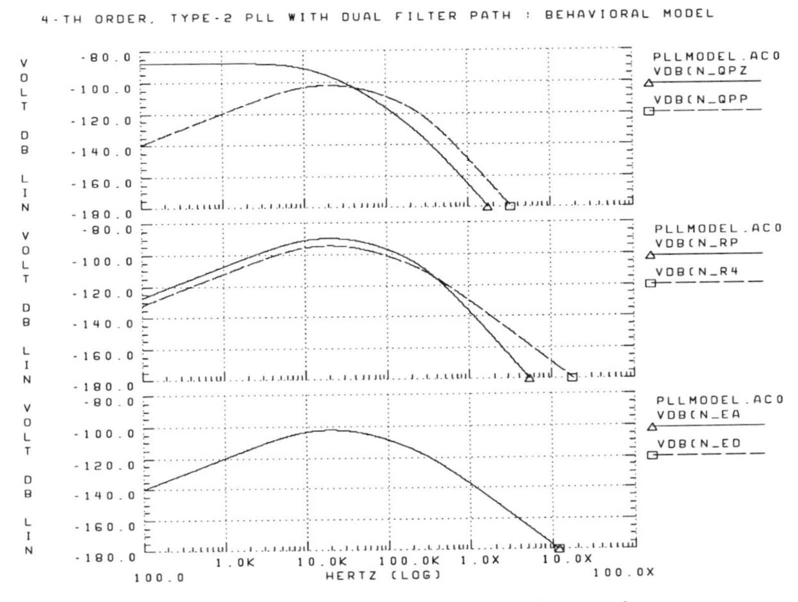

Figure 7.18. Simulated PLL output phase noise

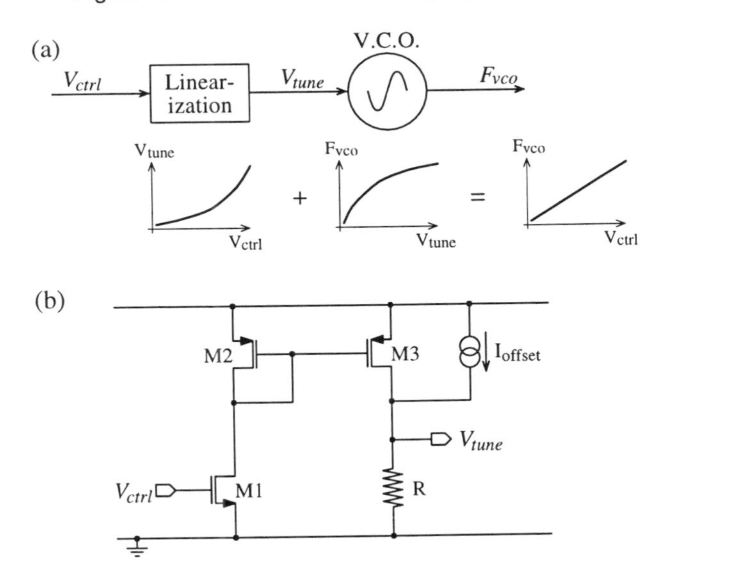

Figure 7.19. Direct linearization of VCO tuning sensitivity : (a) Principle; (b) circuit implementation

sates for the large VCO gain in this operating point. For high input voltages, a high transconductance combines with a low tuning sensitivity to a constant overall gain. Of course, the compensation achieved with this circuit is far from exact, and careful tuning of the circuit parameters is required. An optimal sizing of the input transistor W/L, the resistor value, and the offset current can result in an overall gain that varies from 150 to 250 MHz/V. On top of this, due to variations in the absolute value of the NMOS threshold voltage and transconductance, and the sheet resistance for R, the range we have to design for will be even larger than this.

This large range is too high to accurately control the PLL stability, unless the poles and the zero of the loop gain are spaced further away from the crossover frequency. Since this will decrease the phase noise performance and/or increase the capacitor area, it is definitely desirable to limit the gain variation much more than is achievable with this simple circuit.

Another problem is the phase noise introduced into the loop by this extra active circuit. For frequencies outside the loop bandwidth, noise that enters the loop at this point is not filtered by the loop filter before it appears at the output. It has the same transfer characteristic as the VCO phase noise itself, i.e. it is multiplied by the VCO gain and high-passed before it appears as phase noise at the output. To limit this contribution to e.g. -125 dBc/Hz, several tens of mA of current are required.

The solution taken in this design does not linearize the VCO gain directly, but tries to keep the overall open loop gain constant. The open loop gain is given by (7.18), and from this equation we can notice that it is sufficient to keep the product $I_{qp} \times K_{vco}$ constant in order to maintain stability over the whole range. This is the approach that was already depicted in the overall PLL block diagram of figure 7.2. The VCO tuning voltage is measured, and based on this the charge pump current is set by the linearization block. A 5-step piece-wise linear fitting of the VCO gain is done, and current sources are added/removed from the output current I_{pump}. The VCO gain ranges over a factor 6 from 400 to 70 MHz/V, so the charge pump current ranges from 1 to 6 μA. The charge pump current is increased in 5 logarithmic steps, so for every step it increases by a factor $\sqrt[5]{6} = 1.43$. This results in the following ranges :

V_{ctrl} $[V]$	K_{vco} $[MHz/V]$	I_{qp} $[\mu A]$
< 0.6	400	1.0
< 0.8	285	1.4
< 1.1	200	2.0
< 1.6	140	2.9
< 2.2	95	4.2
< 3.0	70	6.0

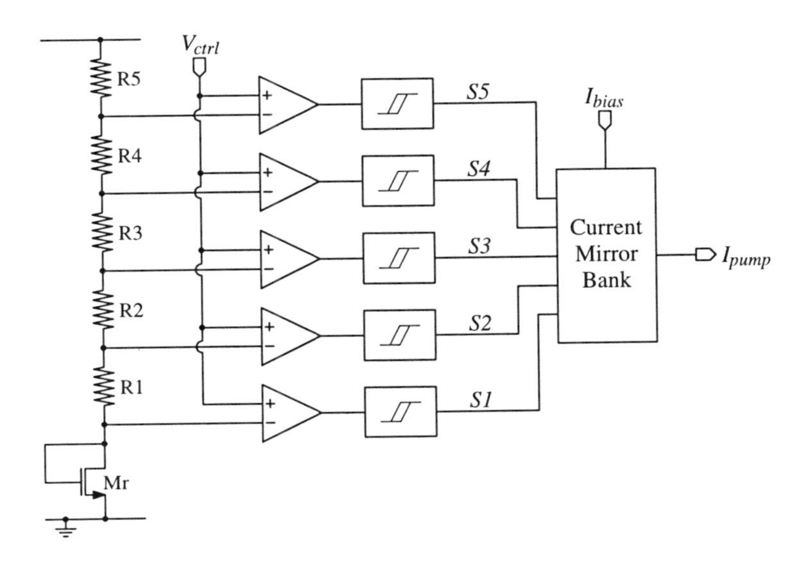

Figure 7.20. Indirect linearization circuitry for the PLL

Figure 7.20 shows the block diagram of the linearization circuit used. The input tuning voltage of the VCO, V_{ctrl}, is compared with 5 reference levels with 5 small amplifiers. These reference levels are generated with the resistor string $R1..R5$ and the transistor Mr. The transistor Mr tracks the absolute value of the NMOS threshold voltage, and hence also tracks the VCO tuning curve, whose absolute position is partly determined by the NMOS amplifying transistors in the oscillator schematic.

The amplifier specs are not severe and the very simple schematic shown in figure 7.21 can be used. A 20-μA bias current is used and the transistor sizes are indicated in the figure. Very small transistors are used. To accommodate for the common mode input range of the amplifier, several NMOS and PMOS source followers shift the voltages of the reference levels and the control voltage up and down. These are not shown in the figure.

The signals obtained at the outputs of the 5 amplifiers can be used to add or remove current sources, but problems might arise when the control voltage almost equals one of the reference voltages. The amplifier cannot decide properly whether the switch must be on or off, and instability might occur when the switch is turned on, the pump current increases, the control voltage increases a little, the switch is turned off, the pump current decreases, the control voltage decreases, etc. Therefore Schmitt-triggers are inserted after each amplifier whose hysteresis avoids situations like this. The Schmitt-trigger schematic is based on [Steya EL86] and is shown in figure 7.22.

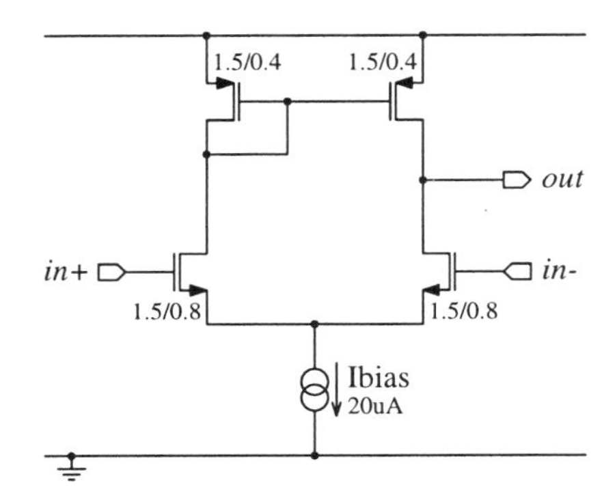

Figure 7.21. Amplifier for indirect linearization circuit

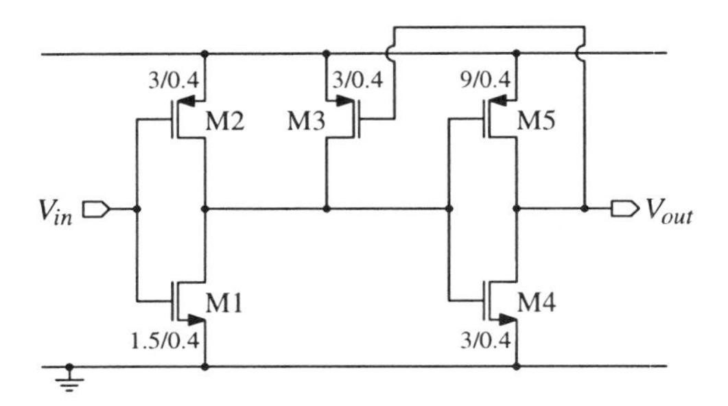

Figure 7.22. Schmitt trigger for indirect linearization circuit

The transistor sizes are indicated in the figure. The two threshold voltages are 1.3 V and 2.3 V. With an amplifier gain of approximately 50, this leaves a sufficiently large value of 20 mV hysteresis on the input control voltage.

The five resulting switch signals $S1..S5$ control each their NMOS transistor which turns a current source on or off, as shown in figure 7.23. The input bias current is supplied externally. Its normal value is $10 \times B \cdot I_{qp} = 120\ \mu A$. It is divided by 10 in an input current mirror, and then used to drive the six mirroring NMOS transistors. The currents through and widths of the transistors in the current mirror bank are listed in the figure. A large gate length of 20 μm has been chosen for the current mirrors, for matching reasons and high output resistance. The switch transistors have a minimal length to limit their resistance. The output current I_{pump} is fed to the charge pump circuitry, where it is used directly for the second pump and divided by twelve for first pump.

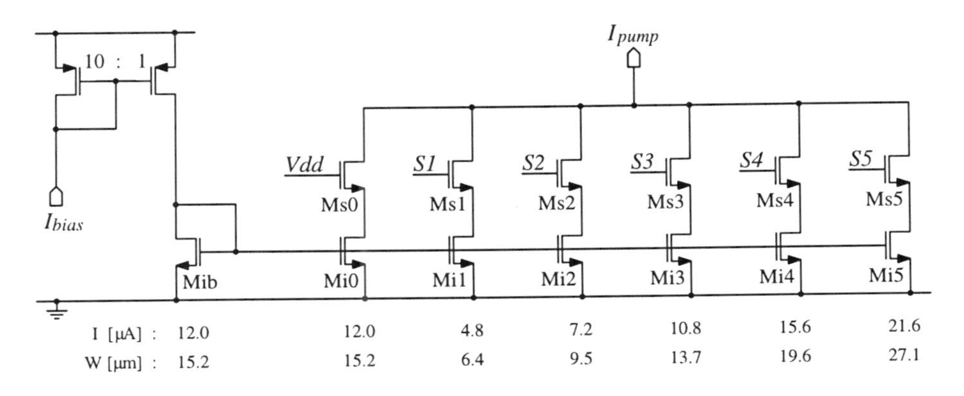

Figure 7.23. Current mirror bank for indirect linearization circuit

We can now get back to the effect of this linearization on the total phase noise. Noise generated in the linearization circuitry is not important, since the pump currents I_{up} and I_{dn} are both deducted from the linearization output current I_{pump}. In the locked state, both the *Up*- and the *Down* current are active for a fraction α_{qp} of the reference signal period, but since the noise that is introduced on these pump currents through I_{pump} is the same for both of them, the total noise currents from I_{Up} and I_{Dn} cancel.

Because of the change in the VCO gain, the transfer function of all the other noise sources to the output phase noise, which has been discussed in section 7.5.2, will also change. As can be seen from the equations (7.26) and (7.28)-(7.31), lowering K_{vco} certainly has a beneficial influence on almost all phase noise contributions. On top of that, as we increase the charge pump current I_{qp} inversely with K_{vco}, an extra improvement is obtained. For example, from table 7.1 we see that the largest amount of phase noise is caused by resistor R_4. Equation (7.31) shows that, if the VCO gain is halved and the pump current is doubled, the phase noise from R_4 decreases with a factor 4, or by 6 dB ! Figure 7.24 shows the variation of the several phase noise parts versus K_{vco}. The charge pump noise falls of with 3 dB per octave, but since this is one of the smaller noise sources, the overall phase noise follows the 6 dB per octave slope of e.g. the noise from R_4.

We can conclude form this figure that we should be able to obtain the required phase noise spec over the largest part of the frequency range.

7.7 MEASUREMENTS

All the circuits described above have been implemented in a single IC shown in figure 7.25. The hollow coil used in the VCO can be clearly distinguished in the lower right hand corner of the die. The prescaler is situated in the lower left hand corner.

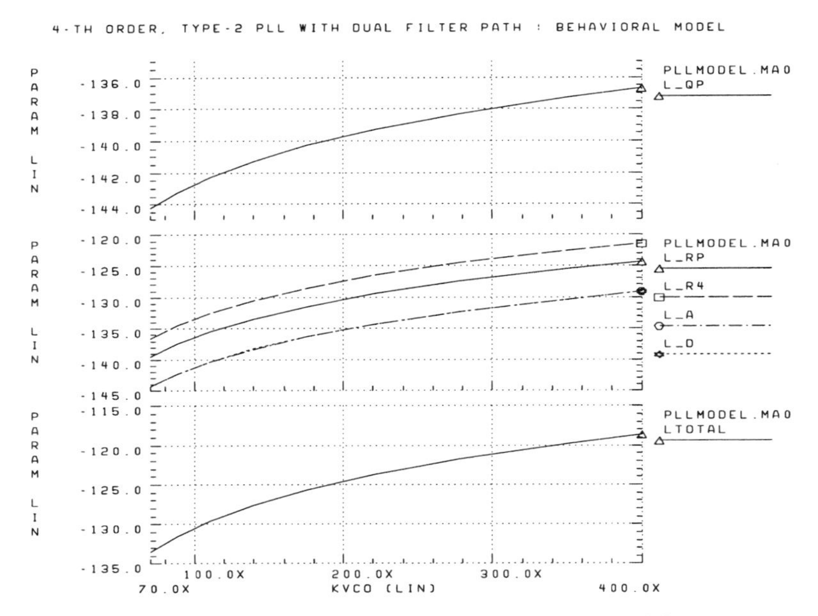

Figure 7.24. Simulated output phase noise versus VCO gain

Both circuits re-use the layout of the previous realizations, except of course for the different placement of the bonding pads. The largest part of the IC is occupied by the three capacitors C_z (320 pF), C_p (160 pF), and C_4 (480 pF). They are realized using the poly/poly structure available in this process with a nominal capacitance value of 1.5 $fF/\mu m^2$. The active filter amplifiers are placed in between these capacitors. The phase-frequency detector and the charge pump are situated in the middle on the left hand side of the die and the linearization circuitry occupies the upper left hand corner. The total die size is 1.7×1.9 μm^2.

The reference frequency is 26.6 MHz and the power supply voltage is 3 V for all circuits. The power consumption of the chip is 51 mW, which is divided over the several building blocks as follows :

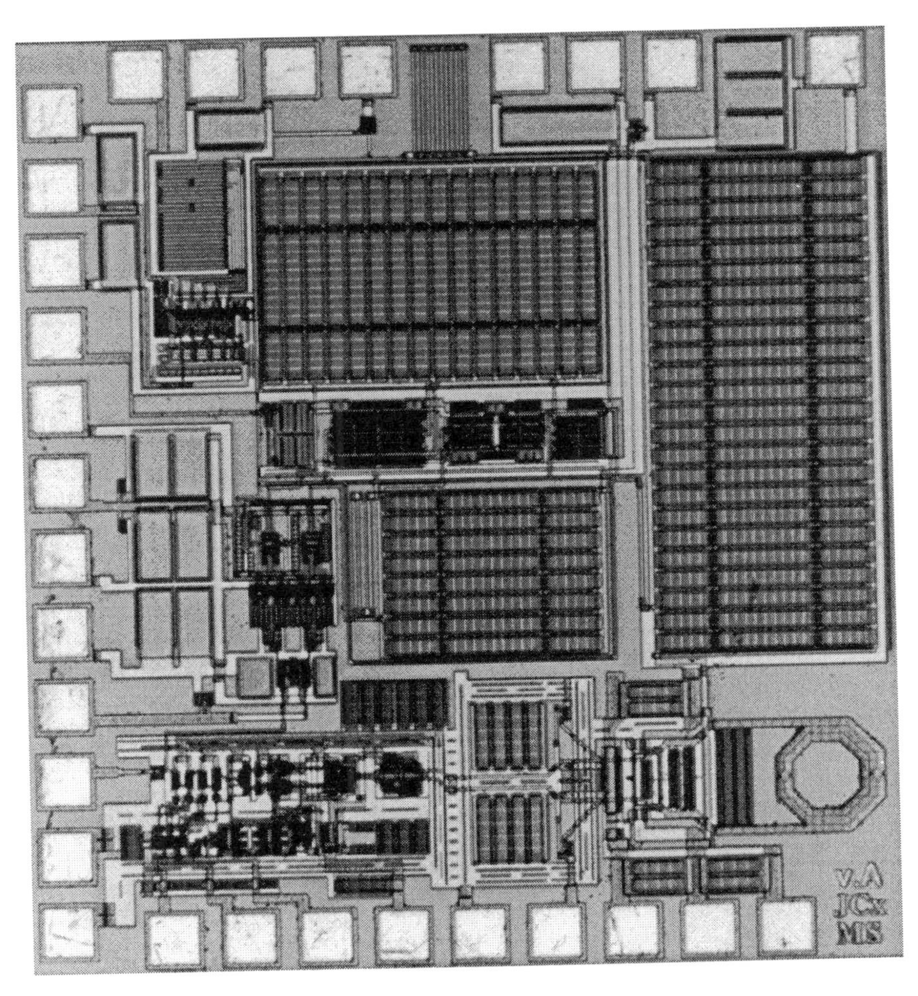

Figure 7.25. IC microphotograph of the complete PLL

PFD	1 mW
Charge Pump	1
Loop filter	18
VCO	11
Prescaler	18
Linearization	2
Total	51 mW

Two important specs to measure are the output phase noise and the transient behavior.

7.7.1 Phase Noise Performance

An important aspect of the variation of the VCO gain with frequency is of course the different noise transfer functions and hence different output phase noise. This is even deteriorated by the increase in $1/\omega^3$ noise in the VCO itself at low tuning voltages. Figure 7.26 shows the output spectrum for two output frequencies, i.e. 1.71 and 1.88 GHz.

Outside the loop bandwidth ($f > 45kHz$), we clearly see the dominant VCO noise. For the 1.71-GHz signal it consists mainly of upconverted $1/f$ noise which results in a magnitude of -112 dBc/Hz at 400 kHz offset. Because of the smaller non-linearities when operating at 1.88 GHz, the VCO spectrum is much cleaner and the PLL phase noise is correspondingly lower, down to -117.5 dBc/Hz at 400 kHz. Extrapolating this to 600 kHz offset results in -121 dBc/Hz, which is sufficient according to the DCS-1800 spec.

Inside the loop bandwidth, it is not the charge pump noise that is dominant, as would be predicted by the simulations performed in figure 7.18. At low offset frequencies, the charge pump feeding the integrater with C_z (QP_z) should give a phase noise contribution of -87 dBc/Hz, which is the largest among all building blocks. But the modeled noise coming from the integrater amplifier and the adder circuit in the bottom part of figure 7.18 takes only the thermal white noise into consideration. It falls of with a slope of 20 dB/dec inside the loop. Unfortunately, missing data on the $1/f$ noise of the transistors in the active filter circuitry caused us to underestimate this contribution to the output phase noise. The phase noise that can be seen in the offset frequency range from 3 to 20 kHz is actually the $1/f$ noise of the biasing circuitry for the amplifier and adder in the active filter. The current sources in these circuits ($M7 - M9$ in figure 7.14 and $M9$ in figure 7.15) are biased with a voltage $V_{b,n}$ for the NMOS and $V_{b,p}$ for the PMOS transistors. These bias voltages are generated from an externally supplied current and some current mirror transistors. It is these transistors

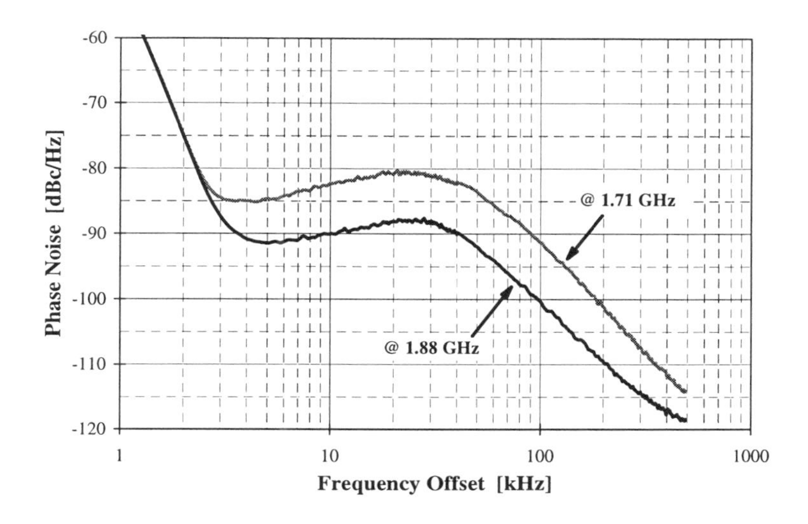

Figure 7.26. Measured PLL output phase noise

that generate a too large amount of $1/f$ noise and become dominant in the in-band phase noise. The $1/f$ increase in the transistor noise magnitude compensates partly for the -20-dB per decade slope in the transfer function, resulting in the measured slope of -10 dB per decade that can be seen in figure 7.26. The highest noise level is measured around 20 kHz offset and reaches a wprst-case level of -80 dBc/Hz for operation at 1.71 GHz. For the 1.88-GHz output, the lower VCO gain constant lowers the in-band phase noise contribution by approximately 8 dB. Closer than 3 kHz from the carrier, the measurement is no longer valid because of the limited accuracy of the spectrum analyser resolution.

It is also this $1/f$ noise that causes the out-of-band PLL phase noise to be a few dB worse than the standalone VCO noise reported in chapter 5. Its magnitude is so high that it is not yet sufficiently suppressed to be neglected with respect to the VCO noise at 400 kHz offset. Again, the rather high noise floor of the measurement setup made it impossible to measure the output spectrum lower than approximately -120 dBc/Hz.

Apart from the close-in phase noise, another important aspect of the output spectrum are the spurs coming from the reference frequency. Although a high value of 26.6-MHz was used for the reference, these spurs still fall inside the DCS-1800 transmit or receive band which is 75 MHz wide. So e.g. in the receive mode the downconverter will also pick up the unwanted frequency spaced 26.6 MHz or 133 channels away. According to the DCS-1800 specifications [ETSI 94], this unwanted signal can have a blocking level of -26 dBm, which is 74 dB higher than the minimum wanted signal power level of -100 dBm. Incorporating a 9-dB margin for sufficient SNR, the spurious level should be lower than -83 dBc.

Coupling of the digital signals in the phase-frequency detector and the charge pump to the VCO through the power supply, and non-complete cancelling of the charge injection in the charge pump switches will modulate the output with a frequency equal to the reference frequency. The charge pump design has been optimized with respect to this aspect, and care has been taken in order to provide sufficient symmetry in the layout. The layout also contains some power supply decoupling capacitors and different bonding pads for sensitive building blocks. The resulting measured spurious suppression is $-75\ dBc$, which is a very good result, but still not sufficient for the DCS-1800 spec.

7.7.2 Transient Characteristics

The second important thing that has been measured is the settling time after a change in the division factor N. Figure 7.27 compares the simulated and measured waveforms of the VCO control voltage when changing the division modulus from 64 to 68 and back. The simulations were done with the mixed-signal simulator Saber [Saber] using digital gates to build the prescaler and the phase-frequency detector. Ideal opamps and current sources were used in the charge pump and the loop filter, and the VCO was represented with an ideal behavioral model. As can be seen in the figure, the simulated control voltage initially increases linearly as long as the input frequency and divider frequency differ significantly. The simulated rise time is approximately 120 μsec. There is a small overshoot, and the voltage settles within an accuracy of 0.25 μV (which corresponds to 100 Hz considering the VCO gain of 400 MHz/V) after a total settling time of 250 μsec.

The measured waveforms differs slightly from the simulated one. The initial slope is not constant but consists of a few linear pieces. This is the result of the piece-wise linearization of the VCO gain through changing the charge pump current. This was not implemented in the behavioral simulations. If we should look at the instantaneous VCO frequency instead of the control voltage, the curve would be linear as well. But that is of course much more difficult to measure.

Apart from this, the measurement agrees very well with the simulations. As the simulation result was obtained with a behavioral model of the PLL, the DC level of the VCO control voltage is not relevant. The slight difference in the rise and fall times can be attributed to an absolute mismatch in e.g. the actual resistor or capacitor values due to process variations. The measurement is not accurate enough to determine the actual settling time to a voltage with μV accuracy, but based on the simulations and the small difference in rise time we can conclude that the PLL has a settling time smaller than 300 μsec.

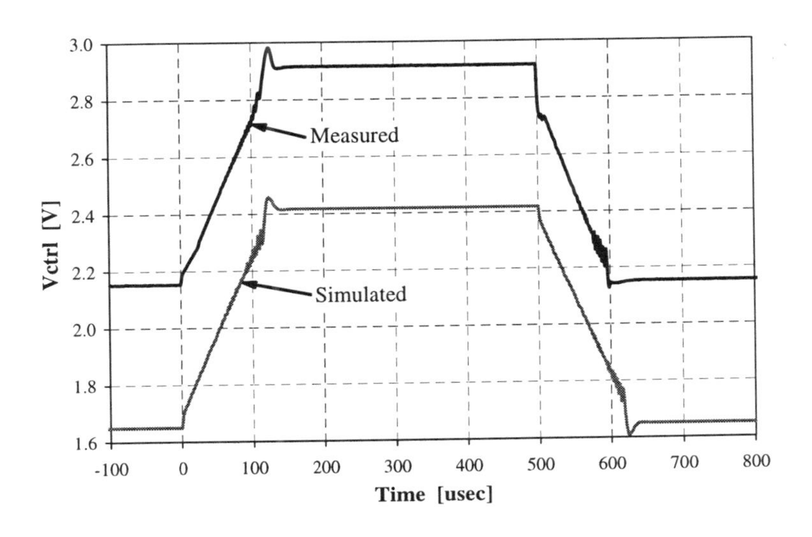

Figure 7.27. Measured PLL settling

7.8 CONCLUSIONS

In this final chapter we have described a working prototype for a DCS-1800 frequency synthesizer. Two approaches taken have had an important influence on the design. First, a 26.6-MHz reference frequency has been chosen, and an eight-modulus prescaler that divides by all integer numbers from 64 to 71 has been used as the single component of the PLL's frequency divider. This combination covers the complete receive and transmit band of the system. Of course, the required 200-kHz frequency resolution is not possible with the prototype design presented here, but must be obtained by implementing the fractional-N division technique, controlled by a delta-sigma modulator to shape the resulting noise appropriately.

The second design aspect has had more consequences on the PLL implementation. We have tried to cover the full frequency band of interest with a simple VCO which has a large enough tuning range. This simplifies the design of the control logic, and avoids the problems involved with the parasitic resistance and capacitances of the transistor switches of the capacitor bank which can degrade the LC-tank's quality factor. But this also means that the VCO is used in its worst case operating point, i.e. at low bias voltages across the tuning capacitor which corresponds to a very high VCO gain constant. This high VCO gain has greatly influenced the design of the several PLL building blocks.

The two high-frequency PLL building blocks, the VCO and the prescaler, have been discussed in chapters 5 and 6. The phase-frequency detector uses a very standard circuit schematic. Most design effort has been put in the charge pump and the loop filter. The charge pump uses eight different control signals to switch on or off the current sources with a clock feedthrough as small as possible. It has been implemented twice to accommodate for the dual loop filter used. This dual loop filter circuit uses two signal paths, one of which uses a scaled current, to create a small time constant without the need for very large capacitor values. A behavioral model of the PLL was implemented in Spice to optimize the loop parameters for low noise. In spite of the optimizations in circuit design and sizing, the resulting design must have a total capacitor value of almost 1 nF to assure the DCS-1800 specs. The loop bandwidth equals 45 kHz. This is mostly a consequence of the large VCO gain, which transfers the noise in the charge pump and the loop filter with a large gain to the output spectrum. Large capacitor values must be used to limit this contribution. A special linearization technique was used for the VCO, because direct linearization of the VCO gain itself leads to excessive power needs to keep the phase noise down. The approach taken here keeps the product of the charge pump current and the VCO gain constant over the complete frequency range, which is also sufficient to create a constant open loop gain, and hence assures stability.

The PLL was realized on a $1.7 \times 1.9\ mm^2$ die in a standard 0.4-μm CMOS process. Of course, it uses no external components and functions without tuning, trimming or post processing steps. It consumes 51 mW from a single 3-V power supply, and is fully functional over the complete frequency range of interest.

The output phase noise complies with the DCS-1800 standard of -119 dBc/Hz at 600 kHz offset, apart from two remaining problems, both of which are related to $1/f$ noise. The first one is the too high ω^{-3} noise of the VCO for the lowest output frequencies. Outside the loop bandwidth, the VCO noise dominates and hence the required spec is not met. This has already been discussed in chapter 5. The second problem is a smaller one, and consists of $1/f$ noise in the biasing circuitry of the active filter amplifiers that has not been sufficiently taken into consideration during the design because of missing data. It causes the in-band phase noise to be slightly larger than expected in some cases, but the level still remains below the required -80 dBc/Hz. In a redesign this can be easily remedied by sizing these biasing transistors appropriately larger, as there are no speed considerations to be taken into account.

For the second important spec of the frequency synthesizer, i.e. the settling time, we clearly benefit from the large reference frequency used. The loop bandwidth is 45 kHz, a number which would not have been possible with a 200-kHz reference. We would have liked to use an even larger bandwidth, but in that case the phase noise coming from the loop components such as the charge pump and the loop filter is not sufficiently suppressed at 600 kHz offset to remain below the system spec. The measured settling time for a 100-MHz frequency step is below 300 μsec, which is fast enough for the DCS-1800 standard.

To conclude, we can state that, although this design does not fulfill the DCS-1800 specification over the complete frequency range, it has demonstrated the feasibility of a full CMOS PLL with no external components as the frequency synthesizer in a modern high-performant mobile communication system such as DCS-1800. As a suggestion for further improvement of the PLL, we think that coarse tuning of the VCO with a binary weighted bank of capacitors should be seriously considered. If one wants to be able to handle 3σ process variation of 20% in the junction capacitor values, this will anyway be necessary as the full VCO tuning range must already be used for this.

Coarse digital tuning with a capacitor bank will have its disadvantages, such as careful design of the switches in the bank and the need for a well-conceived digital control system. But it will enable to operate the VCO in the region with low gain K_{vco}, which will remove the $1/f$ noise problem in the VCO and facilitate the design of the other PLL building blocks. Due to the high VCO gain in this design, the noise transfer from the circuit blocks to the output phase noise has a rather high gain, which has resulted in a maximum value of only 45 kHz for the loop bandwidth, and more important for a very large amount of integrated capacitance. Limiting K_{vco} will probably enable a small increase in the loop bandwidth, but not very much since the noise at 600 kHz offset must be reduced sufficiently by the low-pass function of the loop. But in the loop filter the RC time constants required can be made using a larger (and more noisy) resistor and a smaller capacitance for the same amount of output phase noise. So the required chip area can probably be reduced by a large amount, and the DCS-1800 specs can be achieved over the full frequency band.

8 GENERAL CONCLUSIONS

Now that every home already has its radio, television set, video recorder, personal computer, compact disc player and camcorder, digital mobile phones have become the electronic consumer product of the nineties. The evolution to digital cellular systems has had many consequences. New digital modulation schemes allow a very efficient use of the available spectrum bandwidth, and a large amount of digital signal processing guarantees an excellent speech quality in all conditions. A modern mobile terminal fits in your pocket, weighs less than 200 g, has several hours of talk time and several days of standby time. But most of all, prices have dropped to a level affordable for a large part of the population. The number of applications and subscribers is growing rapidly, and a lot of providers are competing for their share of the market.

The advances in microelectronics have been a determining factor in this evolution. Submicron process technology has resulted in DSP chips with more and more computing power. The RF front end has shifted from a discrete component realization to a solution with only a few ICs. Research is going on investigating the possibility of integrating the complete front end on a single chip. This should enable a large cut in volume, mass, cost and power consumption. The choice of the technology used for this is very important. GaAs has a very good frequency performance, but is also expensive. Bipolar transistors are a good alternative, but integration of large digital processing blocks is not efficient. Therefore BiCMOS seems to give the best results, as this can use bipolar transistors in the high-frequency blocks and MOS transistor for the digital logic. But another trend has risen lately, i.e. the integration in a standard CMOS technology. Deep submicron transistors have f_T values of several tens of GHz, which gives sufficient frequency performance for analog signal processing at $1 - 2\ GHz$, and a powerful digital baseband signal processing core can be added on the same die. It is perhaps not the most economical solution at this time, because of the larger power consumption with respect to a bipolar or GaAs RF front end, but as the MOS gate length scales further it will certainly become so. That is why already today a lot of research is done on the single chip CMOS RF transceiver for mobile communication systems.

Basically, a transceiver consists of three blocks : the receiver, the transmitter, and the LO frequency synthesizer. A lot of design challenges can be found in each block, but this work has tackled the problems of the single chip CMOS frequency synthesizer. The general theory of the indirect or phase-locked loop synthesizer has already been studied in several text books, and the main conclusions have been presented here in chapter 2.

For the VCO, an oscillator based on the resonance of an LC-tank is the only one that gives a sufficiently high output frequency together with a low noise. Chapter 3 has presented a general theory of phase noise in LC-oscillators. A passive inductor with a low series resistance is definitely required, as this is beneficial for low noise as well as low power. Of course, this is exactly the problem one faces when designing an integrated VCO, i.e. high-quality inductors are not readily available in a standard silicon process. With a given inductor of a certain quality, and a maximum voltage swing at the terminals of the active elements, it is not possible to go beyond a certain phase noise level. This can be regarded as a technological limit on the oscillator's phase noise.

The theory developed here describes each LC-tank with an *effective resistance* and *capacitance*, that enable the designer to accurately characterize its performance in terms of phase noise and power consumption. The concept of *enhanced LC-tanks* has been introduced, that allows to make a trade-off between noise and power. They consist of multiple inductors and capacitors that create large internal signals, but maintain normal signal levels at the terminals. These large internal signal levels increase the signal-to-noise ratio and thereby decrease the phase noise. These circuits allow to go beyond the technological noise limit, of course at the expense of an increase in power consumption and area.

Next, two possible implementations of CMOS integrated inductors are evaluated in great detail. Chapter 4 starts off with a more exotic alternative, i.e. inductors based on the parasitic inductance usually associated with a bonding wire. First-order formulas for the calculation of the inductance value are readily available, but the actual value will differ from this because of changes in the height above the substrate, vertical or horizontal bends, or the wire diameter. Evaluation of all these effects results in an expected accuracy of 6% in the inductance value. The inductor's parasitics have also been discussed. They include the bonding pad capacitance and resistance, the wire series resistance with the influence of the skin effect, and losses in the underlying substrate. A demonstrator VCO has been designed, operating at 1.8 GHz in a 0.7-μm CMOS process, using four bonding wires in an enhanced LC-tank. The resulting phase noise of -115 dBc/Hz at an offset of 200 kHz is the lowest noise obtained by any integrated oscillator published in open literature. This result proves the potential of high-quality bonding wire inductors, and the power-noise trade-off with enhanced LC-tanks.

The following chapter deals with the more common known type of integrated inductors, i.e. planar spiral coils laid out in one or more of the standard available metal routing levels. They have already been used in many designs, but in order to cope with their large parasitics most of them rely on changes in the process technology to achieve suitable phase noise specs. Such changes include very thick metal layers, thick field oxides to isolate the inductor from the substrate, or even a complete substrate etch locally underneath the coil. This work has presented a more intelligent approach. An efficient finite-element simulation strategy is proposed that allows to quickly evaluate a spiral coil with a certain geometry. That way all the parasitic effects that appear in the structure can be analyzed, and some qualitative design rules have been proposed to optimize the inductor geometry :

- limit the width of the conductor tracks
- use minimum spacing between the conductors
- use a *hollow* coil shape, omit the innermost conductors
- limit the area of the coil

Two VCOs have been designed which use an optimal hollow coil for the given application, instead of just incorporating a known and characterized coil in the oscillator. The first one uses a standard 0.7-μm CMOS process with a heavily doped substrate. The planar inductors suffer from a lot of substrate losses, and achieves -116 dBc/Hz at an offset of 600 kHz from the 1.8-GHz carrier. Although this is not as good as the bonding wire inductor, which uses the same technology, it is still an order of magnitude better than comparable designs. This improvement must be solely attributed to the use of an optimized hollow coil. A second design profits from the resistive substrate used in a standard 0.4-μm CMOS process. The absence of substrate losses results in planar inductors with high quality factors. A new symmetrical octagonal coil shape is introduced, and the resulting VCO, which operates at 1.8 GHz, achieves -113 dBc/Hz at 200 kHz offset. This is only 2 dB worse than the bonding wire design.

The second high-frequency building block of the PLL, the frequency divider, is discussed in chapter 6. The disadvantages of a fixed-modulus prescaler that has to be inserted when the speed limit of the technology is reached, are overcome with a new dual-modulus prescaler topology, the *phase-switching prescaler* architecture. This new architecture allows to create a dual-modulus divide-by-$N/N+1$ prescaler that has a simple divide-by-4 circuit in its input stage. Its operation is based on the 90-degrees phase shift between the outputs of the master and the slave in a divide-by-2 toggle-flipflop. This divide-by-4 input stage can operate as fast as any asynchronous divider, but yet the full circuit can realize two or even more division factors. This will enable a redesign of existing PLL frequency dividers, resulting in an improvement in the overall PLL specifications by e.g. using a lower division factor, together with a larger reference frequency and a larger loop bandwidth.

A very high speed divide-by-2 master/slave toggle-flipflop is also developed to demonstrate the capabilities of the new prescaler architecture. It is based on an ECL-alike circuit and can operate faster than other existing CMOS circuits. A 0.7-μm CMOS version offers two division factors (128/129) and achieves a maximum input frequency of 1.75 GHz. An eight-modulus prescaler that can divide by all integer numbers from 64 to 71 is realized in a standard 0.4-μm CMOS process, and will enable to cover the complete DCS-1800 frequency band with the prescaler as the sole building block of the frequency divider.

The DCS-1800 system is taken as a demonstrator case for the complete phase-locked loop synthesizer. The 0.4-μm CMOS VCO and the eight-modulus prescaler are combined with the low-frequency components of a PLL to form a complete prototype synthesizer. A 26.6-MHz reference is used, and the loop components are designed to comply with the system's output phase noise specification. The resulting circuit is a 4th-order, type-2 charge pump PLL. Noise requirements limit the loop bandwidth to 45 kHz, and a dual-path loop filter topology is used in order to limit the size of the required capacitors. Nevertheless, a total capacitance of almost 1 nF is integrated on a 1.7 $mm \times$ 1.9 mm die. A special linearization technique is developed that avoids a low-noise (= high-power) direct linearization of the VCO itself by operating on the charge pump current instead of the VCO control voltage. The full circuit consumes 51 mW from a single 3-V supply, and achieve the required DCS-1800 output phase noise, apart from two problems, both of which are related to $1/f$ noise. First, nonlinearities in the VCO at low tuning voltages allow upconversion of baseband $1/f$ noise, which dominates the phase noise outside the loop bandwidth at low output frequencies. Secondly, $1/f$ noise of the loop filter bias circuitry slightly degrades the in-band phase noise. The loop's settling time is only 300 μsec which is certainly sufficient.

Remaining Challenges

With the operational prototype for a DCS-1800 frequency synthesizer we have demonstrated the feasibility of standard CMOS circuits in a mobile terminal RF front end. However, some aspects still require some more attention.

A general problem that will be encountered more and more in deep submicron CMOS circuits is the $1/f$ noise. Both noise problems in the presented synthesizer are an excellent example of that. Smaller gate lengths have of course the advantage of improved frequency performance, but the smaller transistor areas may also result in an increase in the $1/f$ noise. This is of course very dependent on the technology evolution, as $1/f$ noise is also inversely proportional with gate oxide capacitance per area, which increases for the thinner oxides used in submicron transistors. If the $1/f$ noise level increases, and the transistor's corner frequency shifts to values in the range of several MHz, low-noise design techniques will become an important issue in the future. The PLL output phase noise from the active loop filter can easily be reduced by using larger transistors, but in the VCO this is not possible without a reduction of the available tuning range.

Furthermore, the thinner oxides used are also the reason for a larger Spice KF factor because of e.g. more surface scattering. As the $1/f$ noise level increases, and the transistor's corner frequency shifts to values in the range of several MHz, low-noise design techniques will become an important issue in the future. The noise from the active filter can easily be reduced by using larger transistors, but in the VCO this is not possible without a reduction in the available tuning range.

If it is not possible to maintain the VCO tuning range together with low $1/f$ noise, we suggest that it is best if only the highest part of the frequency range is used in the PLL system. This will have several advantages. Due to the highly linear operation of the VCO in that range, the problem of upconverted $1/f$ noise will be negligible. Secondly, the VCO gain will be rather low (e.g. $< 100\ MHz/V$) which will lower the transfer gain of the loop components noise to the output phase noise. As a result, higher resistor values and smaller capacitors can be used in the loop filter. A big design issue in this system is of course the realization of a binary weighted capacitor bank that can be switched in or out to perform a course tuning of the VCO frequency. This adds a lot of parasitics to the oscillator which must not disturb the LC-tank quality. And proper control of the capacitor bank to perform correct frequency acquisition and fast settling in the PLL is also mandatory.

To achieve a 200-kHz frequency resolution starting from a 26.6-MHz reference, a fractional-N divider with sigma-delta modulator is also required. The design of this modulator must also be done with great care, as it will introduce a new source of noise that must be shaped properly to avoid degradation of the output spectrum. The eight-modulus prescaler used perhaps open the way to a multi-bit sigma-delta modulator instead of the standard single bit with a dual-modulus frequency divider. It will be an interesting research topic to see whether this can lead to an improved design.

So some challenges remain open, but yet we have high hopes that the remaining problems can be solved, and a complete PLL LO frequency synthesizer can be realized in a standard CMOS process without any external components or special processing.

Together with the advances in receiver and transmitter circuits, this will enable in the future the realization of a highly integrated, low cost full CMOS transceiver.

Bibliography

[Abidi ACD96] A. A. Abidi, A. Rofougaran, G. Chang, J. Rael, J. Chang, M. Rofougaran, P. Chang, and S. Khorram, "A monolithic 900-MHz spread-spectrum wireless transciever in 1-μm CMOS", in *Analog Circuit Design*, W. M. C. Sansen, J. H. Huijsing, and R. J. van de Plassche, Eds., ISBN 0-7923-9776-2, Kluwer Academic Publishers, Dordrecht, The Netherlands, 1996.

[Abidi CASI92] A. A. Abidi, "Noise in active resonators and the available dynamic range", *IEEE Trans. on Circuits and Systems - I : Fundamental Theory and Applications*, vol. 39, no. 4, pp. 296–299, April 1992.

[Abidi CICC94] A. A. Abidi, "Radio-frequency integrated circuits for portable communications", in *Proc. of the IEEE 1994 Custom Integrated Circuits Conference*, San Diego, USA, May 1994, pp. 151–158.

[Abidi ISSCC97] A. A. Abidi, A. Rofougaran, G. Chang, J. Rael, J. Chang, M. Rofougaran, and P. Chang, "The future of CMOS wireless recievers", in *ISSCC Digest of Technical Papers*, San Fransisco, USA, February 1997, pp. 118–119.

[Abidi JSSC83] A. A. Abidi and R. G. Meyer, "Noise in relaxation oscillators", *IEEE Journal of Solid-State Circuits*, vol. SC-18, no. 6, pp. 794–802, December 1983.

[AD7886] –, "AD7886, a 12-bit, 750-kHz sampling ADC", Data sheet, Analog Devices, April 1991.

[Ali ISSCC96] A. Ali and L. Tham, "A 900-MHz frequency synthesizer with integrated LC voltage-controlled oscillator", in *ISSCC Digest of*

Technical Papers, San Fransisco, USA, February 1996, pp. 390–391.

[Baghd IEEE65] E. J. Baghdady, R. N. Lincoln, and B. D. Nelin, "Short-term frequency stability : Characterization, theory and measurements", *IEEE Proceedings*, vol. 53, pp. 704–722, July 1965.

[Banu ISSCC93] M. Banu and A. Dunlop, "A 660-Mb/s CMOS clock recovery circuit with instantaneous locking for NRZ data and burst-mode transmission", in *ISSCC Digest of Technical Papers*, San Fransisco, USA, February 1993, pp. 102–103.

[Banu JSSC88] M. Banu, "MOS oscillators with multi-decade tuning range and gigahertz maximum speed", *IEEE Journal of Solid-State Circuits*, vol. 23, no. 6, pp. 1386–1393, December 1988.

[Based ESSC94] P. Basedau and Q. Huang, "A 1-GHz, 1.5-V monolithic LC oscillator in 1-μm CMOS", in *Proc. of the 1994 European Solid-State Circuits Conference*, Ulm, September 1994, pp. 172–175.

[Chang EDL93] J. Y.-C. Chang, A. A. Abidi, and M. Gaitan, "Large suspended inductors on silicon and their use in a 2-μm CMOS RF amplifier", *IEEE Electron Device Letters*, vol. 14, no. 4, pp. 246–248, May 1993.

[Chen ISSCC93] D.-L. Chen and R. Waldron, "A single-chip 266-Mb/s CMOS transmitter/receiver for serial data communications", in *ISSCC Digest of Technical Papers*, San Fransisco, USA, February 1993, pp. 100–101.

[Crani CICC97] J. Craninckx and M. Steyaert, "A fully integrated spiral-LC CMOS VCO set with prescaler for GSM and DCS-1800 systems", in *Proc. of the IEEE 1997 Custom Integrated Circuits Conference*, Santa Clara, USA, May 1997, pp. 403–406.

[Crani ISSCC95] J. Craninckx and M. Steyaert, "A 1.8-GHz low-phase-noise voltage controlled oscillator with prescaler", in *ISSCC Digest of Technical Papers*, San Fransisco, USA, February 1995, pp. 266–267.

[Crani JSSC95] J. Craninckx and M. Steyaert, "A 1.8-GHz low-phase-noise voltage controlled oscillator with prescaler", *IEEE Journal of Solid-State Circuits*, vol. 30, no. 12, pp. 1474–1482, December 1995.

[Crani JSSC97] J. Craninckx and M. Steyaert, "A 1.8-GHz low-phase noise CMOS VCO using optimized hollow inductors", *IEEE Journal of Solid-State Circuits*, vol. 32, no. 4, pp. 736–744, May 1997.

[Crani VLSI96] J. Craninckx and M. Steyaert, "A 1.8-GHz low-phase noise spiral-LC CMOS VCO", in *1996 Symposium on VLSI Circuits*, Honolulu, USA, June 1996, pp. 30–31.

[Crols 93] J. Crols, *Integration of Receivers for Phase- and Frequency Modulated Systems*, IWONL Annual Report 1994 (in Dutch), September 1993.

[Crols 97] J. Crols, *Full Integration of Wireless Transceivers*, PhD thesis, Katholieke Universiteit Leuven, Belgium, March 1997.

[Crols VLSI96] J. Crols, P. Kinget, J. Craninckx, and M. Steyaert, "An analytical model for planar inductors on lowly doped silicon substrates for high frequency analog design up to 3 GHz", in *1996 Symposium on VLSI Circuits*, Honolulu, USA, June 1996, pp. 28–29.

[Cutle IEEE66] L. S. Cutler and C. L. Searle, "Some aspects of the theory and measurement of frequency fluctuations in frequency standards", *IEEE Proceedings*, vol. 54, pp. 136–154, February 1966.

[Dauph ISSCC97] L. Dauphinee, M. Copeland, and P. Schvan, "A balanced 1.5-GHz voltage-controlled oscillator with an integrated LC-resonator", in *ISSCC Digest of Technical Papers*, San Fransisco, USA, February 1997, pp. 390–391.

[Desme ESSC97] B. De Smedt and G. Gielen, "Accurate simulation of phase noise oscillators", in *Proc. of the 1997 European Solid-State Circuits Conference*, Southampton, GB, September 1997.

[Dobos CICC94] L. Dobos and B. Jensen, "A versatile monolithic 800-kHz to 800-MHz phase-startable oscillator", in *Proc. of the IEEE 1994 Custom Integrated Circuits Conference*, San Diego, USA, May 1994, pp. 8.4.1–4.

[Dunca CICC95] R. Duncan, K. Martin, and A. Sedra, "A 1-GHz quadrature sinusoidal oscillator", in *Proc. of the IEEE 1995 Custom Integrated Circuits Conference*, Santa Clara, USA, May 1995, pp. 91–94.

[Egan 81] W. F. Egan, *Frequency Synthesis by Phase Lock*, ISBN 0-89464-456-4. J. Wiley & Sons, New York, USA, 1981.

[Enam JSSC90] S. K. Enam and A. A. Abidi, "A 300-MHz CMOS voltage-controlled ring oscillator", *IEEE Journal of Solid-State circuits*, vol. 25, no. 1, pp. 312–315, February 1990.

[Enz AICSP95] C. C. Enz, F. Krummenacher, and E. A. Vittoz, "An analytical MOS transisitor noise model valid in all regions of operation and dedicated to low-voltage and low-current applications", *special issue of the Analog Integrated Circuits and Signal Processing journal on Low-Voltage and Low-Power Design*, vol. 8, no. 1, pp. 83–114, July 1995.

[ETSI 94] –, *European Digital Cellular Telecommunications System (Phase 1); Radio Transmission and Reception; DCS Extension*, ETSI GSM 0505-DCS. European Telecommunications Standard Institute, 1994.

[Forou JSSC95] N. Foroudi and T. A. Kwasniewski, "CMOS high-speed dual-modulus frequency dividers for RF frequency synthesis", *IEEE Journal of Solid-State Circuits*, vol. 30, no. 2, pp. 93–100, February 1995.

[Freem 93] E. M. Freeman, *MagNet 5 User Guide - Using the MagNet Version 5 Package from Infolytica*, Infolytica, London and Montreal, 1993.

[Fujis JSSC93] M. Fujishima, K. Asada, Y. Omura, and K. Izumi, "Low-power 1/2 frequency dividers using 0.1-μm CMOS circuits built with ultrathin SIMOX substrates", *IEEE Journal of Solid-State Circuits*, vol. 28, no. 4, pp. 510–512, April 1993.

[Gardn 79] F. M. Gardner, *Phaselock Techniques*, ISBN 0-471-04294-3. J. Wiley & Sons, New York, USA, 2nd edition, 1979.

[Geppe Spec96] L. Geppert, "Technology 1996 : Solid state", *IEEE Spectrum*, pp. 51–55, January 1996.

[Gray 93] P. R. Gray and R. G. Meyer, *Analysis and Design of Analog Integrated Circuits*, ISBN 0-471-59984-0. J. Wiley & Sons, New York, USA, 3rd edition, 1993.

[Grebe 84] A. Grebene, *Bipolar and MOS Analog Integrated Circuit Design*, ISBN 0-471-08529-4. J. Wiley & Sons, New York, USA, 1984.

[Green TPHP74] H. M. Greenhouse, "Design of planar rectangular microelectronic inductors", *IEEE Trans. on Parts, Hybrids and Packaging*, vol. PHP-10, no. 2, pp. 1001–109, June 1974.

[Haigh 89] D. Haigh and J. Everard (ed.), *GaAs Technology and its Impact on Circuits and Systems*, ISBN 0-86341-187-8,. Peter Peregrinus, London, GB, 1989.

[Heine ISSCC97] S. Heinen, K. Hadjizada, U. Matter, W. Geppert, T. Volker S. Weber, S. Beyer, J. Fenk, and E. Matschke, "A 2.7-V 2.5-GHz bipolar chipset for digital wireless communication", in *ISSCC Digest of Technical Papers*, San Fransisco, USA, February 1997, pp. 306–307.

[Hill RFD92] A. Hill and J. Surber, "The PLL dead zone and how to avoid it", *RF Design*, pp. 131–134, March 1992.

[HSpice96] –, *HSpice User Manual*, Meta-Software, Inc., Campbell, USA, 1996.

[Huang JSSC88] Q. Huang, W. Sansen, M. Steyaert, and P. Van Peteghem, "Design and implementation of a CMOS VCXO for FM stereo decoders", *IEEE Journal of Solid-State Circuits*, vol. 23, no. 3, pp. 784–793, June 1988.

[Huang JSSC96] Q. Huang and R. Rogenmoser, "Speed optimization of edge-triggered CMOS circuits for gigahertz single-phase clocks", *IEEE Journal of Solid-State Circuits*, vol. 31, no. 3, pp. 456–465, March 1996.

[Irie ISSCC97] K. Irie, H. Matsui, T. endo, K. Watanabe, T. Yamawaki, M. Kokubo, and J. Hildersley, "A 2.7-V GSM RF transceiver IC", in *ISSCC Digest of Technical Papers*, San Fransisco, USA, February 1997, pp. 302–303.

[Janse ISSCC97] B. Jansen, K. Negus, and D. Lee, "Silicon bipolar VCO family for 1.1 to 2.2 GHz with fully integrated tank and tuning circuits", in *ISSCC Digest of Technical Papers*, San Fransisco, USA, February 1997, pp. 392–393.

[Ji-Ren JSSC89] Y. Ji-Ren and C. Svensson, "High-speed CMOS circuit technique", *IEEE Journal of Solid-State Circuits*, vol. 24, no. 1, pp. 62–70, February 1989.

[Kado JSSC93] Y. Kado, M. Suzuki, K. Koike, Y. Omura, and K. Izumi, "A 1-GHz / 0.9-mW CMOS/SOMIX divide-by-128/129 dual modulus prescaler using a divide-by-2/3 synchronous counter", *IEEE Journal of Solid-State Circuits*, vol. 28, no. 4, pp. 513–517, April 1993.

[Kato JSSC88] K Kato, T. Sase, H. Sato, I. Ikushima, and S. Kojima, "A low-power 128-MHz VCO for monolithic PLL IC's", *IEEE Journal of Solid-State Circuits*, vol. 23, no. 2, pp. 474–479, April 1988.

[Kwasn CICC95] T. Kwasniewski, M. Abou-Seido, A. Bouchet, F. Gaussorgues, and J. Zimmerman, "Inductorless oscilator design for personal communication devices - a 1.2-μm CMOS process case study", in *Proc. of the IEEE 1995 Custom Integrated Circuits Conference*, San Diego, USA, May 1995, pp. 327–330.

[Laker 94] K. R. Laker and W. M. C. Sansen, *Design of Analog Integrated Circuits and Systems*, ISBN 0-07-036060-X. McGraw-Hill, New York, USA, 1994.

[Laund UWRF92] A. Laundrie, "Crystal oscillators continue to set stability standards", *Microwaves & RF*, pp. 140–147, December 1992.

[Leeso IEEE66] D. B. Leeson, "A simple model of feedback oscillator noise spectrum", *IEEE Proceedings*, vol. 54, pp. 329–330, February 1966.

[Liu JSSC90] T.-P. Liu and R. G. Meyer, "A 250-MHz monolithic voltage-controlled oscillator with low temperatur coefficient", *IEEE Journal of Solid-State Circuits*, vol. 25, no. 2, pp. 555–561, April 1990.

[Long JSSC97] J. R. Long and M. A. Copeland, "The modeling, characterization, and design of monolithic inductors for silicon RF IC's", *IEEE Journal of Solid-State Circuits*, vol. 32, no. 3, pp. 357–369, March 1997.

[Marqu JWN97] A. Marques, M. Steyaert, and W. Sansen, "Theory of PLL fractional-N frequency synthesizers", *Journal of Wireless Networks, Special issue: VLSI in Wireless Networks*, vol. X, no. X, pp. xx–xx, Month 1997.

[MC4044] –, "MC4044", Data sheet, Motorola, –.

[Meyr 90] H. Meyr and G. Ascheid, *Synchronization in Digital Communications, vol. 1 : Phase-, Frequency Locked Loops and Amplitude Control*, ISBN 0-471-50193-X. J. Wiley & Sons, New York, USA, 1990.

[Mietec C07] Alcatel Mietec, "Electrical parameters CMOS 0.7 μm".

[Mijus JSSC94] D. Mijuskovic, M. J. Bayer, T. F. Chomicz, N. K. Garg, F. James, P. W. McEntarfer, and J. A. Porter, "Cell based fully integrated

CMOS frequency synthesizers", *IEEE Journal of Solid-State Circuits*, vol. 29, no. 3, pp. 271–279, March 1994.

[Mille TIAM91] B. Miller and R. J. Conley, "A multiple modulator fractional divider", *IEEE Trans. on Instrumentation and Measurement*, vol. 40, no. 3, pp. 578–583, June 1991.

[Min CICC94] J. Min, A. Rofougaran, H. Samueli, and A. Abidi, "An all-CMOS architecture for a low-power frequency-hopped 900-MHz spread spectrum transceiver", in *Proc. of the IEEE 1994 Custom Integrated Circuits Conference*, San Diego, USA, May 1994, pp. 379–382.

[Muerl 94] J. Meurling and R. Jeans, *The Mobile Phone Book*, ISBN 0-9524031-0-2. Communicationsweek International, London, GB, 1994.

[Nguye JSSC90] N. M. Nguyen and R. G. Meyer, "Si IC-compatible inductors and LC passive filters", *IEEE Journal of Solid-State Circuits*, vol. 25, no. 4, pp. 1028–1031, August 1990.

[Nguye JSSC92] N. M. Nguyen and R. G. Meyer, "A 1.8-GHz monolithic LC voltage-controlled oscillator", *IEEE Journal of Solid-State Circuits*, vol. 27, no. 3, pp. 444–450, March 1992.

[Nikne CICC97] A. M. Niknejad and R. G. Meyer, "Analysis and optimization of monolithic inductors and transformers for RF ICs", in *Proc. of the IEEE 1997 Custom Integrated Circuits Conference*, Santa Clara, USA, May 1997, pp. 375–378.

[Nordh CAS90] E. H. Nordholt and C. A. M. Boon, "Single-pin integrated crystal oscillators", *IEEE Transactions on Circuits and Systems*, vol. 37, no. 2, pp. 175–182, February 1990.

[Olgaa NORC93] C. Olgaard and A. Rofourgan, "A low power 900-MHz tuned CMOS amplifier with large output swing capability", in *Proc. of the eleventh Norchip seminar*, Trondheim, Norway, November 1993, pp. 162–169.

[Parke CICC97] J. Parker and D. Ray, "A low-noise 1.6-ghz CMOS PLL with on-chip loop filter", in *Proc. of the IEEE 1997 Custom Integrated Circuits Conference*, Santa Clara, USA, May 1997, pp. 407–410.

[Pottb JSSC94] A. Pottbacker and U. Langmann, "An 8-GHz silicon bipolar clock recovery and data-regenerator IC", *IEEE Journal of Solid-State Circuits*, vol. 29, no. 12, pp. 1572–1578, December 1994.

[Razav CICC97] B. Razavi, "Challenges in the design of frequency synthesizers for wireless applications", in *Proc. of the IEEE 1997 Custom Integrated Circuits Conference*, Santa Clara, USA, May 1997, pp. 395–402.

[Razav ISSCC94a] B. Razavi, K. F. Lee, and R.-H. Yan, "A 13.4-GHz CMOS frequency divider", in *ISSCC Digest of Technical Papers*, San Fransisco, USA, February 1994, pp. 176–177.

[Razav ISSCC94b] B. Razavi and J. Sung, "A 6-GHz 60-mW BiCMOS phase locked loop with 2-V supply", in *ISSCC Digest of Technical Papers*, San Fransisco, USA, February 1994, pp. 114–115.

[Razav ISSCC97] B. Razavi, "A 1.8-GHz CMOS voltage-controlled oscillator", in *ISSCC Digest of Technical Papers*, San Fransisco, USA, February 1997, pp. 388–389.

[Razav JSSC95] B. Razavi, K. F. Lee, and R.-H. Yan, "Design of high-speed, low-power frequency dividers and phase-locked loops in deep submicron CMOS", *IEEE Journal of Solid-State Circuits*, vol. 30, no. 2, pp. 101–109, February 1995.

[Razav JSSC96] B. Razavi, "A study of phase noise in CMOS oscillators", *IEEE Journal of Solid-State Circuits*, vol. 31, no. 3, pp. 331–343, March 1996.

[Riley JSSC93] T. A. D. Riley, M. A. Copeland, and T. A. Kwasniewski, "Delta-sigma modulation in fractional-N frequency synthesis", *IEEE Journal of Solid-State Circuits*, vol. 28, no. 5, pp. 553–559, May 1993.

[Rodri CAS90] A. Rodriguez-Vazques, B. Linares-Barranco, J. L. Huertas, and E. Sanchez-Ninencio, "On the design of voltage-controlled oscillators using OTAs", *IEEE Transactions on Circuits and Systems*, vol. 37, no. 2, pp. 198–211, February 1990.

[Rofou ISSCC96] A. Rofourgan, J. Rael, M. Rofourgan, and A. Abidi, "A 900-MHz CMOS LC-oscillator with quadrature outputs", in *ISSCC Digest of Technical Papers*, San Fransisco, USA, February 1996, pp. 392–393.

[Rogen CICC93] R. Rogenmoser, N. Felber, Q. Huang, and W. Fichtner, "A 1.16-GHz dual modulus 1.2-μm CMOS prescaler", in *Proc. of the IEEE 1993 Custom Integrated Circuits Conference*, San Diego, USA, May 1993, pp. 27.6.1–4.

[Rogen CICC94] R. Rogenmoser, Q. Huang, and F. Piazza, "1.57-GHz asynchronous and 1.4-GHz dual modulus 1.2-μm CMOS prescalers", in *Proc. of the IEEE 1994 Custom Integrated Circuits Conference*, San Diego, USA, May 1994, pp. 16.3.1–4.

[Rudel ISSCC97] J. C. Rudell, J.-J. Ou, T. B. Cho, G. Chien, F. Brianti, J. A. Weldon, and P. R. Gray, "A 1.9-GHz wide-band IF double conversion CMOS integrated receiver for cordless telephone applications", in *ISSCC Digest of Technical Papers*, San Fransisco, USA, February 1997, pp. 304–305.

[Saber] –, *Saber User Manual*, Analogy, 1996.

[Sanse ESSC87] W. Sansen, Q. Huang, M. Steyaert, and P. Van Peteghem, "A CMOS VCXO for FM stereo decoders", in *Proc. of the 1987 European Solid-State Circuits Conference*, 1987, pp. 156–160.

[Santo JSSC84] J. T. Santos and R. G. Meyer, "A one-pin crystal oscillator for VLSI circuits", *IEEE Journal of Solid-State Circuits*, vol. SC-19, no. 2, pp. 228–236, April 1984.

[Sato ISSCC96] H. Sato, K. Kashiwagi, K. Niwano, T. Iga, T.Ikeda, and K. Mashiko, "A 1.9-GHz single-chip IF transceiver for digital cordless phones", in *ISSCC Digest of Technical Papers*, San Fransisco, USA, February 1996, pp. 342–343.

[Senan EL89a] R. Senani and B. Amit Kumar, "Linearly tunable Wien bridge oscillator realized with operational transconductance amplifiers", *IEE Electronic Letters*, vol. 25, no. 1, pp. 19–21, 5th January 1989.

[Senan EL89b] R. Senani and B. Amit Kumar, "New electronically tunable OTA-C sinusoidal oscillator", *IEE Electronic Letters*, vol. 25, no. 4, pp. 286–287, 16th February 1989.

[Shaha ISSCC97] A. R. Shahani, D. K. Shaeffer, and T. H. Lee, "A 12-mW wide dynamic range CMOS front-end for a portable GPS receiver", in *ISSCC Digest of Technical Papers*, San Fransisco, USA, February 1997, pp. 368–369.

[Sneep JSSC90] J. G. Sneep and C. J. M. Verhoeven, "A new low-noise 100-MHz balanced relaxation oscillator", *IEEE Journal of Solid-State Circuits*, vol. 25, no. 3, pp. 692–698, June 1990.

[Soyue JSSC89] M. Soyuer and R. G. Meyer, "High-frequency phase-locked loops in monolithic bipolar technology", *IEEE Journal of Solid-State Circuits*, vol. 24, no. 3, pp. 787–795, June 1989.

[Soyue JSSC96a] M. Soyuer, K. A. Jenkins, J. N. Burghartz, H. A. Ainspan, F. J. Canora, S. Ponnapalli, J. F. Ewen, and W. E. Pence, "A 2.4-GHz silicon bipolar oscillator with integrated resonator", *IEEE Journal of Solid-State Circuits*, vol. 31, no. 2, pp. 268–270, February 1996.

[Soyue JSSC96b] M. Soyuer, K. A. Jenkins, J. N. Burghartz, and M. D. Hulvey, "A 3-V 4-GHz voltage-controilled oscillator with integrated resonator", *IEEE Journal of Solid-State Circuits*, vol. 31, no. 12, pp. 2042–2045, December 1996.

[Steya EL86] M. Steyaert and W. Sansen, "Novel CMOS schmitt trigger", *IEE Electronic Letters*, vol. 22, no. 4, pp. 203–204, 13th February 1986.

[Steya EL94] M. Steyaert and J. Craninckx, "1.1-GHz oscillator using bond-wire inductance", *IEE Electronic Letters*, vol. 30, no. 3, pp. 244–245, 3rd February 1994.

[Thams CICC95] M. Thamsirianunt and T. Kwasniewski, "CMOS VCOs for PLL frequency synthesis in GHz digital mobile radio communications", in *Proc. of the IEEE 1995 Custom Integrated Circuits Conference*, San Diego, USA, May 1995, pp. 331–334.

[vdTan ISSCC97] J. van der Tang and D. Kasperdovitz, "A 0.9 − 2.2-GHz monolithic quadrature mixer oscillator for direct-conversion satellite receivers", in *ISSCC Digest of Technical Papers*, San Fransisco, USA, February 1997, pp. 88–89.

[Vitto JSSC88] E. Vittoz, M. Degrauwe, and S. Bitz, "High-performance crystal oscillators : Theory and application", *IEEE Journal of Solid-State Circuits*, vol. 23, no. 3, pp. 774–783, June 1988.

[Voorm 1993] J. O. Voorman, "Continuous-time analog integrated filters", in *Integrated Continuous-Time Filters*, Y. P. Tsividis and J. O. Voorman, Eds., pp. 27–29. IEEE Press, New York, USA, 1993.

[Wang EL92] Z.-G. Wang, "Multigigahertz varactorless Si bipolar VCO IC", *IEE Electronic Letters*, vol. 28, no. 6, pp. 548–549, 12th March 1992.

[Wang ISSCC94] Z.-G. Wang, M. Berroth, J. Seibel, P. Hofmann, A. Hulsmann, K. Kohler, G. Raynor, and J. Schneider, "19-GHz monolithic integrated clock recovery using PLL and 0.3-μm gate-length quantum-well HEMTs", in *ISSCC Digest of Technical Papers*, San Fransisco, USA, Februari 1994, pp. 118–119.

[Wang JSSC90] Y.-T. Wang and A. A. Abidi, "CMOS active filter design at very high frequencies", *IEEE Journal of Solid-state Circuits*, vol. 25, no. 6, pp. 1562–1574, December 1990.

[Weiga ISCAS94] T. C. Weigandt, B. Kim, and P. R. Gray, "Analysis of timing jitter in CMOS ring oscillators", in *Proc. of the IEEE 1994 International Symposium on Circuits and Systems*, London, GB, May 1994, pp. 27–30.

[Zhang UWGW93] G. F. Zhang and J. L. Gautier, "Broad-band, lossless monolithic microwave active floating inductor", *IEEE Microwave and Guided Wave Letters*, vol. 3, no. 4, pp. 98–100, April 1993.

Index